SAFFRON

Medicinal and Aromatic Plants – Industrial Profiles

Individual volumes in this series provide both industry and academia with in-depth coverage of one major medicinal or aromatic plant of industrial importance.

Edited by Dr Roland Hardman

Volume 1
Valerian
edited by Peter J. Houghton

Volume 2
Perilla
edited by He-Ci Yu, Kenichi Kosuna and Megumi Haga

Volume 3
Poppy
edited by Jenő Bernáth

Volume 4
Cannabis
edited by David T. Brown

Volume 5
Neem
H.S. Puri

Volume 6
Ergot
edited by Vladimír Křen and Ladislav Cvak

Volume 7
Caraway
edited by Éva Németh

Volume 8
Saffron
edited by Moshe Negbi

Other volumes in preparation

Allium, edited by K. Chan
Artemisia, edited by C. Wright
Basil, edited by R. Hiltunen and Y. Holm
Cardamom, edited by P.N. Ravindran and K.J. Madusoodanan
Chamomile, edited by R. Franke and H. Schilcher
Cinnamon and Cassia, edited by P.N. Ravindran and S. Ravindran
Colchicum, edited by V. Šimánek
Curcuma, edited by B.A. Nagasampagi and A.P. Purohit
Eucalyptus, edited by J. Coppen

Please see the back of this book for other volumes in preparation in Medicinal and Aromatic Plants – Industrial Profiles

SAFFRON

Crocus sativus L.

Edited by

Moshe Negbi

Faculty of Agriculture, Food and Environmental Quality Sciences
The Hebrew University of Jerusalem, Rehovot, Israel

harwood academic publishers
Australia • Canada • China • France • Germany • India • Japan
Luxembourg • Malaysia • The Netherlands • Russia • Singapore
Switzerland

Copyright © 1999 OPA (Overseas Publishers Association) N.V. Published by license under the Harwood Academic Publishers imprint, part of the Gordon and Breach Publishing Group.

All rights reserved.

No part of this book may be reproduced or utilized in any form or by any means, electronic or mechanical, including photocopying and recording, or by any information storage or retrieval system, without permission in writing from the publisher. Printed in Singapore.

Amsteldijk 166
1st Floor
1079 LH Amsterdam
The Netherlands

British Library Cataloguing in Publication Data

Saffron : crocus sativus L. – (Medicinal and aromatic
    plants : industrial profiles; v. 8)
    1. Saffron crocus   2. Saffron crocus – Industrial applications
    3. Saffron crocus – Therapeutic use
    I. Negbi, Moshe
    584.3'8

    ISBN 90-5702-394-6
    ISSN 1027-4502

The illustration on the cover is taken from Theophrastus' *De Historia Plantarum*, Amsterdam apud H. Laurentium (typis J. Broerssen), 1664 (page 667, Crocus). Courtesy of Hunt Institute for Botanical Documentation, Carnegie Mellon University, Pittsburgh, PA, USA.

CONTENTS

Preface to the Series   vii
Preface   ix
Contributors   xi

1   Saffron Cultivation: Past, Present and Future Prospects   1
      Moshe Negbi

The Saffron Plant (*Crocus sativus* L.) and its Allies

2   Botany, Taxonomy and Cytology of *C. sativus* L. and its Allies   19
      Brian Mathew

3   Reproduction Biology of Saffron and its Allies   31
      Maria Grilli Caiola

4   Saffron Chemistry   45
      Dov Basker

The Present State of Saffron Cultivation and Technology

5   Saffron (*Crocus sativus* L.) in Italy   53
      Fernando Tammaro

6   Saffron Cultivation in Azerbaijan   63
      N.Sh. Azizbekova and E.L. Milyaeva

7   Saffron Cultivation in Greece   73
      Apostolos H. Goliaris

8   Saffron Cultivation in Morocco   87
      Ahmed Ait-Oubahou and Mohamed El-Otmani

9   Saffron Technology   95
      Dov Basker

Futuristic Aspects of Saffron Cultivation, Usage and Industry

10   Saffron in Biological and Medical Research   103
      Fikrat I. Abdullaev and Gerald D. Frenkel

11 Mechanized Saffron Cultivation, Including Harvesting 115
    Pier Francesco Galigani and Francesco Garbati Pegna

12 Sterility and Perspectives for Genetic Improvement of *Crocus sativus* L. 127
    Giusseppe Chichiriccò

13 *In Vitro* Propagation and Secondary Metabolite Production in *Crocus sativus* L. 137
    Ora Plessner and Meira Ziv

Index   149

PREFACE TO THE SERIES

There is increasing interest in industry, academia and the health sciences in medicinal and aromatic plants. In passing from plant production to the eventual product used by the public, many sciences are involved. This series brings together information which is currently scattered through an ever increasing number of journals. Each volume gives an in-depth look at one plant genus, about which an area specialist has assembled information ranging from the production of the plant to market trends and quality control.

Many industries are involved such as forestry, agriculture, chemical, food, flavour, beverage, pharmaceutical, cosmetic and fragrance. The plant raw materials are roots, rhizomes, bulbs, leaves, stems, barks, wood, flowers, fruits and seeds. These yield gums, resins, essential (volatile) oils, fixed oils, waxes, juices, extracts and spices for medicinal and aromatic purposes. All these commodities are traded world-wide. A dealer's market report for an item may say "Drought in the country of origin has forced up prices".

Natural products do not mean safe products and account of this has to be taken by the above industries, which are subject to regulation. For example, a number of plants which are approved for use in medicine must not be used in cosmetic products.

The assessment of safe to use starts with the harvested plant material which has to comply with an official monograph. This may require absence of, or prescribed limits of, radioactive material, heavy metals, aflatoxins, pesticide residue, as well as the required level of active principle. This analytical control is costly and tends to exclude small batches of plant material. Large scale contracted mechanised cultivation with designated seed or plantlets is now preferable.

Today, plant selection is not only for the yield of active principle, but for the plant's ability to overcome disease, climatic stress and the hazards caused by mankind. Such methods as *in vitro* fertilisation, meristem cultures and somatic embryogenesis are used. The transfer of sections of DNA is giving rise to controversy in the case of some end-uses of the plant material.

Some suppliers of plant raw material are now able to certify that they are supplying organically-farmed medicinal plants, herbs and spices. The Economic Union directive (CVO/EU No 2092/91) details the specifications for the **obligatory** quality controls to be carried out at all stages of production and processing of organic products.

Fascinating plant folklore and ethnopharmacology leads to medicinal potential. Examples are the muscle relaxants based on the arrow poison, curare, from species of *Chondrodendron*, and the antimalarials derived from species of *Cinchona* and *Artemisia*. The methods of detection of pharmacological activity have become increasingly reliable and specific, frequently involving enzymes in bioassays and avoiding the use of laboratory animals. By using bioassay linked fractionation of crude plant juices or extracts, compounds can be specifically targeted which, for example, inhibit blood platelet aggregation, or have antitumour, or antiviral, or any other required activity. With the assistance of robotic devices, all the members of a genus may be readily screened. However, the plant material must be **fully** authenticated by a specialist.

The medicinal traditions of ancient civilisations such as those of China and India have a large armamentarium of plants in their pharmacopoeias which are used throughout South East Asia. A similar situation exists in Africa and South America. Thus, a very high percentage of the world's population relies on medicinal and aromatic plants for their medicine. Western medicine is also responding. Already in Germany all medical practitioners have to pass an examination in phytotherapy before being allowed to practise. It is noticeable that throughout Europe and the USA, medical, pharmacy and health related schools are increasingly offering training in phytotherapy.

Multinational pharmaceutical companies have become less enamoured of the single compound magic bullet cure. The high costs of such ventures and the endless competition from me too compounds from rival companies often discourage the attempt. Independent phytomedicine companies have been very strong in Germany. However, by the end of 1995, eleven (almost all) had been acquired by the multi-national pharmaceutical firms, acknowledging the lay public's growing demand for phytomedicines in the Western World.

The business of dietary supplements in the Western World has expanded from the Health Store to the pharmacy. Alternative medicine includes plant based products. Appropriate measures to ensure the quality, safety and efficacy of these either already exist or are being answered by greater legislative control by such bodies as the Food and Drug Administration of the USA and the recently created European Agency for the Evaluation of Medicinal Products, based in London.

In the USA, the Dietary Supplement and Health Education Act of 1994 recognised the class of phytotherapeutic agents derived from medicinal and aromatic plants. Furthermore, under public pressure, the US Congress set up an Office of Alternative Medicine and this office in 1994 assisted the filing of several Investigational New Drug (IND) applications, required for clinical trials of some Chinese herbal preparations. The significance of these applications was that each Chinese preparation involved several plants and yet was handled as a single IND. A demonstration of the contribution to efficacy, of each ingredient of each plant, was not required. This was a major step forward towards more sensible regulations in regard to phytomedicines.

My thanks are due to the staff of Harwood Academic Publishers who have made this series possible and especially to the volume editors and their chapter contributors for the authoritative information.

Roland Hardman

PREFACE

The saffron plant (*Crocus sativus* L.) and its spice – the stigmas taken away from saffron flowers and dried – are shrouded in a glorious and expensive air and some ignorance, even among scholars. Being aware of the lack of comprehensive books on saffron I accepted the invitation of Harwood Academic Publishers to edit the current book.

The current volume lacks articles on saffron's cultivation in Spain and India. Unfortunately, appeals to saffronologists in these countries did not produce any results. Contrarily, almost all other requests to contributors resulted in the return of a manuscript. It is with gratitude that I thank these authors for their co-operation, and moreover, for their positive responses to requests for clarifications and amendments. Since this is a multi-authored volume, some points are repeated in several chapters. This is unavoidable, unless the editor serves as a super-author. Likewise in some chapters there are internal repetitions; removing these would have changed the spirit of the articles, thus here they are given as the authors thought best to present them.

This volume is divided into three main sections. The first details the botany, including Brian Mathew's masterful taxonomic chapter, followed by chapters on saffron's reproduction biology and its chemistry. The second deals with the present state of saffron cultivation in Azerbaijan (first published in the West), Greece, Italy and Morocco. The last chapter of the second part is dedicated to saffron technology. The chapters in the last section describe futuristic aspects of saffron cultivation, usage and industry. In this section Pier Francesco Galigani and Francesco Garbati Pegna sum up the studies on saffron technology carried out in Florence, which seems to be the world centre of saffron engineering.

I expect that this volume will evoke enthusiasm for saffron, which I know is positively infectious. It happened to my colleagues and students Naza Azizbekova, Dov Basker, Benjamin Dagan, Ada Dror, Dalia Greenberg-Kaslasi, Ora Plessner, Betsy Sachs and Meira Ziv. Their share in the present volume is larger than the number of times they are cited.

I am grateful to Dr Roland Hardman, the editor of the series, for the invitation to edit this volume and for his continuing advice. I warmly acknowledge the fruitful collaboration of Harwood Academic Publishers and Camille Vainstein of the Faculty of Agriculture at The Hebrew University of Jerusalem. The preparation of this volume was partially supported by the Israel Science Foundation (founded by the Israel Academy of Science and Humanities). I would like to express my gratitude to the Wellcome Institute for the History of Medicine in London where I worked on this volume while on sabbatical leave.

Moshe Negbi

CONTRIBUTORS

Fikrat I. Abdullaev
Unidad de Investigacion en Salud Ifantil
Instituto Nacional de Pediatria
Av. Insurgentes Sur
3700-C 04530 Mexico

Ahmed Ait-Oubahou
Department of Horticulture
Institut Agronomique et Vétérinaire
    Hassan II
B.P. 121
Aït Melloul
80150 Agadir
Morocco

N.Sh. Azizbekova
Plant Science Department
University of British Columbia
Vancouver, B.C.
Canada V6T 1Z4

Dov Basker
Department of Food Science
Agricultural Research Organization
The Volcani Centre
Bet Dagan
Israel

Maria Grilli Caiola
Department of Biology
University of Rome "Tor Vergata"
Via delle Ricerca Scientifica 1
00133 Rome
Italy

Giusseppe Chichiriccò
Department of Environmental Sciences
University of L'Aquila
Via Vetoio
67100 L'Aquila
Italy

Mohamed El-Otmani
Department of Horticulture
Institut Agronomique et Vétérinaire
    Hassan II
B.P. 121
Aït Melloul
80150 Agadir
Morocco

Gerald D. Frenkel
Department of Biological Sciences
Rutgers University
101 Warren Street
Newark
NJ 07102
USA

Pier Francesco Galigani
Dipartimento di Ingegneria Agraria e
    Forestale
Università degli Studi di Firenze
Piaza delle Cascine
15-50144 Firenze
Italy

Apostolos H. Goliaris
Department of Aromatic and Medicinal
    Plants
Agricultural Research Centre
    of Macedonia-Thrace
57001 Thermi-Thessaloniki
Greece

Brian Mathew
90 Foley Road
Claygate
KT10 0NB
UK

E.L. Milyaeva
Timiryazev Institute of Plant Physiology
Russian Academy of Sciences
35 Botanicheskaya Street
Moscow
127276 Russia

Moshe Negbi
Department of Agricultural Botany
Faculty of Agriculture, Food and
 Environmental Quality Sciences
The Hebrew University of Jerusalem
P.O. Box 12
Rehovot 76100
Israel

Francesco Garbati Pegna
Dipartimento di Ingegneria Agraria e
 Forestale
Università degli Studi di Firenze
Piaza delle Cascine
15-50144 Firenze
Italy

Ora Plessner
Department of Agricultural Botany
Faculty of Agriculture, Food and
 Environmental Quality Sciences
The Hebrew University of Jerusalem
P.O. Box 12
Rehovot 76100
Israel

Fernando Tammaro
Department of Environmental Sciences
University of L'Aquila
Via Vetoio
67100 L'Aquila
Italy

Meira Ziv
Department of Agricultural Botany
Faculty of Agriculture, Food and
 Environmental Quality Sciences
The Hebrew University of Jerusalem
P.O. Box 12
Rehovot 76100
Israel

1. SAFFRON CULTIVATION: PAST, PRESENT AND FUTURE PROSPECTS

MOSHE NEGBI

Department of Agricultural Botany,
The Hebrew University of Jerusalem,
PO Box 12, Rehovot,
76100, Israel

ABSTRACT   This is an introductory chapter to a volume on saffron, containing 12 more chapters ranging from saffron biology, chemistry, cultivation, and new developments in agronomic methods to novel uses. This article deals briefly with the domestication of the saffron crocus (*Crocus sativus* L.), describes the present state of saffron cultivation around the world, and gives acreage and production when available. The recent developments in cultivation and production methods, in the field and *in vitro*, are discussed. Traditional uses of saffron and the development of new uses in histochemistry and medicine are reviewed.

"Nothing is adulterated as much as saffron," Pliny (NH 21 32).

INTRODUCTION

Evidence from Aegean Bronze Age art, Linear B documents, classical authors (Theophrastus, 371–287 BC, Pliny, 23–79 AD and Dioscorides, 1st century AD) and taxo-cytological studies was brought together in an attempt to outline the process of saffron domestication. In this study, Negbi and Negbi (in press) argued that saffron was first harvested from the wild *Crocus cartwrightianus*, a mutant of which – *Crocus sativus*, distinguished by its elongated stigmas – was observed, selected and domesticated on Crete during the Late Bronze Age. Saffron was later established as a minor but expensive crop in the Old World from India to Britain (Warburg 1957). Nonetheless, the detailed history of saffron's establishment in various regions of the Old World awaits a categorical study. During the past four decades, saffron's glorious past has been praised and its present difficulties lamented (Amigues 1988, Anon. Undated, Basker and Negbi 1983, Coppock 1984, Di Francesco 1990, Fois Sussarello 1990, Greenberg and Lambert Ortiz 1983, Ingram 1969, Jossen and Stork 1983, Mir 1992, Negbi and Negbi, in press, Raines Ward 1988, Rees 1988, Skrubis 1990, Szita 1987, Tammaro 1990, Tammaro and Di Francesco 1978, Tuveri 1990).

Saffron cultivation has been linked with either concealed or overt research in traditional countries of cultivation such as Azerbaijan, France, Greece, India, Iran, Italy and Spain, and in other countries such as China, Israel, Japan and Mexico.

Progress in cultivation methods in Italy, Greece and Israel was presented at two recent symposia (Bezzi 1987, Tammaro and Marra 1990), and is described in an agricultural manual (Tammaro and Di Francesco 1978, Di Francesco 1990) and in this volume (Basker, Chichirricò, Galigani and Garbati Pegna, Grilli Caiola, Goliaris, Tammaro and Plessner and Ziv). Research in India is published periodically (Bali and Sagwal, 1987, Chrungoo and Farooq 1984, Dhar *et al.* 1988, Koul and Farooq 1984, Nair *et al.* 1992, Nauriyal *et al.* 1977).

Moreover, recently there has been a tendency to outline future technological breakthroughs (Chichiriccò 1990, Laneri 1990, Negbi 1990). The envisaged Utopia, which would enable continued saffron production, is based on research on the following:

(A) genetic improvement, via breeding the saffron *crocus* (*C. sativus*) with its close allies,
(B) developments in corm production *in vitro*,
(C) *in vitro* techniques to cultivate the spice itself, namely the stigmas of the saffron flower.

The present volume comes a decade after Tammaro and Marra's proceedings "Lo zafferano" (1990), in a period when many saffronologists are exploring ways of improving saffron cultivation and generating the spice. Several chapters are devoted to these futuristic aspects, and in all the other chapters, better methods of saffron production are contemplated.

SAFFRON CULTIVATION AREA, PRODUCTION AND EXPORT

The dried stigmas of the saffron *crocus*, in their natural form, as coupe saffron and powder, are produced worldwide at an annual rate of 50 tonnes (Oberdieck 1991). Spain is the world's foremost exporter, and perhaps producer, of saffron. A USDA Foreign Agricultural Service report summarizing Spanish production and export is presented here (Table 1.1), showing that Spain has retained approximately the same acreage for three decades (Ingram 1969). It is worth noting that the average yield has

Table 1.1 Data from Spain on cultivation area, production and exports. USDA foreign agricultural service's data (Anon. 1992)

Period (Yearly Average)	Area (ha)	Production (kg)	Yield[1] (kg per ha)	Export (Tonnes)
1960–64	4,987	42,242	8.47	23
1965–69	4,244	29,865	7.84	25
1970–74	4,334	47,543	10.97	36
1975–79	4,477	36,314	8.11	26
1980–84	4,050	28,763	7.10	27
1985–88	4,184	29,153	6.97	34[2]

[1] Average yield for the whole period is 8.24 kg per ha.
[2] Export data available only for the years 1985 and 1986.

remained at 7 to 11 kg per ha, values much higher than those obtained in most other countries, except Italy (and Turkey?). The largest importers of saffron are Saudi Arabia and the Gulf Emirates, with the USA coming next (Raines Ward 1988). Saffron imports to the USA in recent years have averaged over 3 tonnes annually and been valued at about $3,300,000. Virtually all shipments have been from Spain, with most of the balance coming from Italy and India (Anon. 1992). The same source gives a New York spot price of $1,045 per kg for Spanish saffron, in each of the first three months of 1992. Retail prices in the USA have risen to $5,000 per kg and up to $8,000 per kg at Bloomingdale's, and even to $14,000 in 0.25 g packets at Macy's (Allan and Fowler 1985, Raines Ward 1988, Szita 1986). Prices of raw Spanish saffron in Germany fluctuate between DM 1,200 and 2,100 per kg (Oberdieck 1991). In Britain, retail prices of £4,500 per kg have been recorded (Rees 1988).

Saffron is cultivated in Greece by a 2,500-family-strong co-operative in Krokos and neighbouring Macedonian villages (Skrubis 1990). During the 1980s, there was a dramatic reduction in saffron cultivation in Greece (Table 1.2). The reduction in acreage and yield may reflect a common cause for abandoning saffron cultivation, namely increasing labour costs. Even in the best years, the yields in Krokos were lower than the Spanish average, and this is probably not solely due to the relative drought conditions, as stated by Skrubis (1990).[1] Greece exports most of its saffron to Germany, Switzerland, China, Sweden, the UK, the USA and Hong Kong, and even to Italy and Spain. From 1979 to 1989, the income varied from $800 to $1,000 per kg. In Southern Greece, Crete, Santorini and other Aegean islands, saffron is still being collected from the wild *C. cartwrightianus* (Mathew 1982, pers. commun. September 1997). Saffron production in Morocco is confined to areas not frequently described in the literature (Chitt *et al.* 1985, Wallach 1989). In the Rivers of Palms irrigation system, in Morocco, saffron, henna and rose (for rose oil) are cultivated as cash crops and from a common 2-acre plot, a family earns about $350 annually. An updated report on saffron in Morocco by Ait-Oubahou and El-Otmani appears in this volume.

Available data on world production of saffron is presented in Table 1.3. I could not find production data for some countries known to grow saffron, such as Austria,

Table 1.2 Saffron production in Krokos and adjacent villages, Macedonia, Greece from 1982 to 1988 (based on Skrubis 1990)

Year	Area (ha)	Production (kg)	Yield[1] (kg per ha)
1982	1,600	12,500	7.8
1983	1,200	8,408	7.0
1984	1,100	6,030	5.5
1985	1,000	2,095	2.1
1986	900	960	1.1
1987	880	2,710	3.1
1988	860	3,690	4.3

[1] Average yield for the whole period is 4.4 kg per ha.

Libya and Mexico (Di Francesco 1990, Löpez Camacho and Fucikovsky 1985). Saffron is cultivated in the semiarid Khorasan region of Iran (Behzad *et al.* 1992a,b). According to a newspaper article, some 6,000 ha in this region produce 30 tonnes annually (Anon. 1989). However, further documentation is needed to confirm that Iran is the largest producer of saffron. In India, saffron cultivation is for the most part limited to a small region in Pampore, Kashmir (Mir 1992). For the latest available figures on the cultivation area and yield see Siddique *et al.* (1989).[2]

Italy produces saffron in Altopiano di Navelli (Provincia dell Aquila), Sardinia[3] and Emilia-Romagna. During the 13th century, saffron was grown in Tuscany and even exported to the Levant (Abulafia, 1882). Recently, cultivation in San Giminiano, Tuscany has been started (Galigani and Garbati Pegna, this volume). The highest saffron yields have been reported from Navelli, where they vary from 10 to 18 kg of dried stigmas per ha (Tammaro 1990). Over the past 160 years, cultivation areas in Navelli have decreased, from 450 ha in 1830, 300 ha in 1910, and under 10 ha in the 1980s, to 6 ha in 1992. Production over the years has fluctuated greatly, dropping from 5.5 tonnes to 1.4 kg, then rising again to 90 kg in 1992 (Di Crecchio 1960, Di Francesco 1990, and pers. commun., Tammaro 1990, Tammaro and Di Francesco

Table 1.3 Available data on world production of saffron

Country	Area (ha)	Production (kg)	Yield (kg per ha)	Source
Iran	6,000	30,000	5	Anon. (1989)
Spain	4,184	29,153		Anon. (1992)
India (a)	2,440	4,800	2	Nauriyal *et al.* (1977)
India (b)	240	1,200	5	Siddique *et al.* (1989)
Greece	860	3,698	4.3	Skrubis (1990)
Azerbaijan	675	–	–	Azizbekova & Milyaeva (1999)
Morocco	500	1,000	2–2.5	Ait-Oubahou & El-Otmani (1999)
Italy:	29.4	–	–	Marzi (1990)
Italy by regions:				
Sardinia	19	160	8.42	Tuveri (1990)
Naveli	6	90	15	Di Francesco[1] (1993)
Emilia-Romagna	0.1	–	–	Zanzucchi[2] (1993)
France	3	–	–	Aucante[3] (1989)
Turkey	0.1	2	20?	Vurdu[4] (1993)
Switzerland	–	0.4	–	Jossen & Stork (1983)

[1] L. Di Francesco, Ispettorato Provinciale dell'Agricoltura, L´Aquila, pers. commun.
[2] C. Zanzucchi, Consorzio Cumunale Parmensi, pers. commun.
[3] Saffron is mainly grown in the Orleans area. French acreage and export is not well cited: Ingram (1969) cites exports in 1966 as 1 tonne, Bali and Sagwal (1987) cited acreage and export (for 1972/73) as 400 ha and 1 tonne, respectively. Brighton (1977) and Daniel Royer (pers. commun.) do not give any details on saffron cultivation in France.
[4] Dr H. Vurdu, Middle East Technical University, Ankara, pers. commun.

1978). Sardinia is the largest producer in Italy and the area of saffron cultivation there is increasing. Fourteen out of 19 ha are cultivated near San Giovano Monreale and the yields are from 8 to 9 kg per ha (Tuveri 1990).

Saffron production in Turkey is limited to the town of Safranbolu near the Black Sea, and was quite low in the early 1980s (Mathew 1984). A few years ago, only one farmer grew it, on an area covering a little over 0.1 ha, with maximum yields of 20 kg per ha (H. Vurdu, pers. commun.).

Reductions in saffron cultivation areas in some major producer countries have not discouraged efforts to revive its cultivation and even grow it in new territories. Cultivation experiments carried out in Japan (Kawatani *et al.* 1961) appear to have been discontinued, though corm exports are reported from that country (Szita 1987). However, related studies are still in progress, especially *in vitro* culture for corms and stigma production (Himeno and Sano 1987, Himeno *et al.* 1988, Isa and Ogasawara 1988, Kamikura and Nakazato 1984, Koyama *et al.* 1988, Otsuka *et al.* 1992, Sano and Himeno 1987, Sarma *et al.* 1990, 1991).

In China, cultivation trials have been carried out in Peking, Changchun and Zhejiang provinces (Chen and Sze 1977, Lu *et al.* 1988, Yang and Miao 1985). The 1985–87 trials in Zhejiang yielded very small amounts of the spice – 0.8 kg of dry flowers obtained from a mu (0.067 ha), i.e. 11.9 kg per ha. This would be a good yield if the translator had written "dry stigmas" instead of "dry flowers". Saffron is probably also produced in Tibet (Szita 1987). Experimental studies in Israel have been reported (Negbi *et al.* 1989, Negbi 1990). Other experiments have been initiated on a semi-industrial scale (N. Sh. Azizbekova, Y. Foa and E. Shiloni, pers. commun.).

The major producers of saffron do not engage in corm export (Szita 1987). In Navelli, under-sized corms are fed to pigs (pers. commun.). However, cultivation areas are extended using locally grown saffron corms in India (Madan *et al.* 1966) and probably in other countries. The Netherlands and Japan produce and export corms (Szita 1987), but the extent of this trade is unknown to me.

PHYSIOLOGICAL STUDIES

Flowering

Control of flowering has been studied recently in an attempt to enhance hysteranthous (following flowering) leaf appearance in the saffron crocus. This normally sub-hysteranthous geophyte (Mathew 1982) flowers in autumn, before (as in Macedonia, Goliaris, this volume), concomitant with or following leaf appearance. Treated plants flowered without previous root or leaf emergence, thus hysteranthously, if the corms were kept dry at 15°C (Plessner *et al.* 1989). This study was performed to enable the replacement of labour-intensive manual picking of the flowers for spice separation with a mechanical harvester prior to planting. Late corm planting did not harm corm production (Plessner *et al.* 1989, compare with Kawatani *et al.* 1961).[4]

In the genus *Crocus*, the flower is situated on a minute peduncle (Dafni *et al.* 1981, Fritsch 1939, 1942, Mathew 1982, Wilkins 1985). Only the elongated perianth tube connects the tip of the peduncle with the free perianth lobes, which are barely above ground. The peduncle elongates in fertile crocus species after fertilization and eventually puts the capsules at ground level. In an English translation of a Russian article by Azizbekova *et al.* (1978), one gets the impression that treatment with gibberellin induced elongation of the flower stalk. In fact, it represented only an increase in the length of the perianth tube. Hormonal treatment that would cause elongation of the true flower stalk (peduncle) at flowering is therefore still needed.[5]

Flower ontogenesis of *C. sativus* was studied in Azerbaijan and Russia (Azizbekova and Milyaeva 1979, Milyaeva and Azizbekova 1978), Kashmir (Koul and Farooq 1984) and Israel (Greenberg-Kaslasi 1991). Although the anatomical details are quite similar in the three studies, the timing of the transition from vegetative to reproductive shoot apex differed. In Azerbaijan, it began in March, in Kashmir in July and in Israel in March–April. Flower organs formed in July, with almost complete flowers being observed by the end of the month. Greenberg-Kaslasi (1991) suggested that corm size or seasonal variations determine the difference in transition dates. She proposed that the exact internal and environmental variables determining transition period and flower development are essential to the timing of flowering.

Corm Production

The importance of adequate corm production is self-evident in the sterile taxon saffron crocus, which has been reproduced vegetatively for millennia by annually replacing corms. This practice was described as early as 300 BC by Theophrastus in *Historia Plantarum* 6.6.10 and in the first century AD by Pliny in *Natural History* 21.32 (Negbi 1989). Since almost every sprouting bud produces a corm and there are about 10 buds on a flowering-size corm, factors affecting sprouting are highly important for corm production. The size of the daughter corms is equally important, since for the most part, only flowering-size corms are used for planting. The relationships between corm size, flower number and weight of stigmatic lobes have recently been described (De Mastro and Ruta 1993).

Planting depth affects corm production: more buds sprouted from shallowly planted corms than from deeply planted ones – resulting in more daughter corms (Negbi 1990). Corms treated with gibberellins$_{4+7}$ before planting had a decreased number of sprouting buds, resulting in fewer daughter corms, although the apical one grew to a larger size (Greenberg-Kaslasi 1991). This finding is explained by increased apical dominance.

Thus effective production of daughter corms results from a combination of shallow planting and dominance of the apical bud. In any event, *in vitro* corm production remains a desirable goal, at least to attain virus-free corms (Dhar and Sapru 1993 (1994), George *et al.* 1992, Homes *et al.* 1987, Isa and Ogasawara 1988, Plessner *et al.* 1990, Milyaeva *et al.* 1995). Plessner and Ziv updated this subject for the present volume.

Root Development and Function

Three kinds of roots are distinguished in *C. sativus* (Figure 1 in Negbi 1990):

(1) Absorbing roots that emerge from the base of the planted corm: they are thin and relatively long.

(2) Contractile roots, thick and short, that develop singly at the base of the sprouting buds which form daughter corms. They are produced only in shallowly located corms. Though their main function is to deepen the newly formed corms, they absorb water and nutrients.
(3) Contractile-absorbing roots which are thinner and longer than the contractile ones and develop on the parent corm near the sprouting buds bearing the contractile roots.

The contractile-absorbing roots appear three weeks after the contractile ones (Greenberg-Kaslasi 1991). This strengthens the hypothesis that their formation is induced not only by depth factors such as light and fluctuating temperatures (Halevy 1986, in *Gladiolus*), but also by the presence of the contractile roots. While contractile roots decrease to half or a third of their length, the thin-contractile roots contract by only 10%. Although the function of the thin-contractile roots needs further elucidation, they may serve to anchor the mother corm in position during the tilting activity of the contractile roots.

STERILITY AND IMPROVEMENT OF THE SAFFRON CROCUS

Mathew (1977, 1982), Brighton (1977) and Chichiriccò (1984) studied the taxonomy, morphology and cytology of *C. sativus* and its allies. *C. sativus* is a sterile triploid (2n=24); it produces no fertilizable gametes[6] (Ghaffari 1986) and is self-incompatible (Chichiriccò and Grilli Caiola 1986, 1987). In the rare cases of fertilization there is some embryo and endosperm development, processes that terminate at early stages (Chichiriccò 1987). Studies of micro- and macro-sporogenesis and pollen-tube growth have shown that *C. sativus* is essentially a male-sterile plant, due mainly to pollen malformation and malfunction (Chichiriccò 1989a,b, 1990, Chichiriccò and Grilli Caiola 1986, 1987). Recently, a study of pollen grains, pollen tube and pistil was carried out in three members of the *C. sativus* group: *C. thomasii*, *C. cartwrightianus* and *C. biflorus* Miller subsp. *biflorus* (Grilli Caiola *et al.* 1993, Grilli Caiola, 1994, 1995; Chichiriccò and Grilli Caiola have contributed chapters on progress in the field to this volume).

Three fertile diploid (2n=16) *Crocus* species are considered as putative ancestors of the sterile saffron crocus: the Aegean *C. cartwrightianus*, the Cretan *C. oreocreticus* (Mathew 1982) and the Italian and Dalmatian *C. thomasii* (Chichiriccò 1990). Of these, crosses with the saffron crocus have been attempted only with the last. Chichiriccò (1989a, 1990) managed to achieve viable hybrids between *C. sativus* × *C. thomasii*. The continuation and extension of this study to include the Aegean allies of the saffron crocus is of theoretical and applied significance (Grilli Caiola 1995, see Grilli Caiola's article in this volume.[7]) Another member of the saffron crocus group, *C. moabiticus* (2n=14), has been recently rediscovered, described, brought from Jordan and placed in crocus collections in Europe (Al-Eisawi 1986, Kerndorf 1988). It may also add to future breeding programmes.

Studies in the 1960s and 1970s dealt with the selection of rare saffron *Crocus* plants in the field, such as plants with more than three stigmas and other such traits. Estilai (1978) screened large fields for similar qualities, but no further reports were published. In Israel, plants with supernumerary stigmas were marked, but these traits did not recur the following year. Similar freaks with 4 and 5-fid stigmas, which were observed at a frequency of about one in a million, did not reappear (Dhar et al. 1988).

Another approach, increasing variability for selection by irradiation, has not received much attention. Gamma-irradiation (up to 0.5 Kr) of saffron corms increased corm production, flower number and stigma weight for three consecutive years (Akhund-Zade and Muzaferova 1975). An unpublished study in L'Aquila (U. Laneri, S. Lucretti and F. Tammaro 1983) showed many morphological flower variants following 12 Gy gamma-irradiation of hundreds of saffron corms. However, no useful mutants were recovered, and moreover, further flowering was inhibited (Laneri 1990). There is a need to continue this approach.

DRYING AND STORAGE OF SAFFRON AND QUALITY DETERMINATION

Following the separation of the stigmas from the flowers, it is essential to dry them. Incomplete drying results in total loss of the product due to decomposition and moulding. Dried, uncontaminated stigmas are storable and marketable, but may not be of the highest quality, as determined by colouring power (crocin concentration), odour (safranal) and taste (picrocrocin) (Basker 1993). Some modern drying methods have replaced the age-old drying on fine-mesh screens held over burning coals (Anon. Undated, Raines Ward 1988). Controlled storage of the harvested stigmas is used in a number of producing countries (Skrubis 1990, Zanzucchi 1987).

In Krokos, Greece, a thin layer of freshly harvested stigmas is placed on a fine silk screen and dried in a dark, oven-warmed room for 10 to 12 h, at temperatures reaching 30–35°C. A red spice of similarly high quality has been obtained experimentally in a dehydrating chamber at 48°C for 3 h. There they found that storage of the dried spice was best achieved in tightly sealed glass jars or cans, in 10–12% RH. Saffron analysis in Greece carried out according to the French norm[8] was based on a determination of moisture and volatile-material content, colouring power, pigment identification, and ash, impurity, and volatile-oil contents (Skrubis 1990).

In the L'Aquila region, Italy, the author witnessed drying over hot charcoal two decades ago. It seems that these and similar skilful and delicate methods are still in use in Emilia-Romagna, over oak wood amber (Zanzucchi 1987), and in Navelli (Di Francesco 1990, Tammaro this volume). However, analytical methods similar to those used in Greece are employed in Emilia-Romagna (Zanzucchi 1987). Saffron determination in Navelli also includes nitrogenous and non-nitrogenous substances, sugars, cellulose and oils (Tammaro 1990: 76–79). More detailed spectroscopic determinations of the spice's ingredients were performed by Cichelli (1990) using chromatographic techniques (GLC and HPLC). A new HPLC method, adapted for this purpose, has been recently published (Corti et al. 1996).

In Spain, to my knowledge, the actual drying methods employed have not been published (but it was dried over glowing coals at least until the early 1980s, Greenberg and Lambert Ortiz 1983: 76–78). However, analytical methods used to study the auto-oxidation of saffron have been described. Under the summer temperature and humidity of La Mancha (40°C and 75% RH), loss in colouring power and bitterness follow first- and second-order kinetics, respectively (Alõnso *et al.* 1990, see also Iborra *et al.* 1992 and Alõnso *et al.* 1996). These authors raised the possibility of using their techniques to monitor the characteristics and purity of market saffron.

A recent, detailed study by Basker (1993) describes the direct relationship between the drying temperature of the stigmas and the time needed for effective results. Drying at temperatures ranging from 20° to 92°C for 96 to 1.2 h, respectively, resulted in similar amounts of crocin (colouring power, expressed as Extinction$^{1\% \, w/v}$ on a path of 1 cm at 440 nm), picrocrocin (bitterness factor), safranal (aroma factor) and sensory quality (Basker and Negbi 1985). Basker has updated these topics in two contributions to this volume.

USES OF SAFFRON

Apiculture

C. sativus and other crocuses are pollinated by the honeybee (*Apis mellifera*), solitary bees and syrphids flies (Ferrazzi 1991, Shmida and Dafni 1989). *A. mellifera* gathers the pollen from these flowers in big orange pellets. Nectar is produced from the septa of the ovary, placed at the base of the long and narrow corolla tube near ground level. Nectar does not always reach the perianth to be sucked up by short-tongued insects such as honeybees. Various *Crocus* species represent a good resource for honeybee colonies during periods of hardship: between winter and spring and in autumn, but are usually unimportant in terms of honey production in Italy (Ferrazzi 1991).

Traditional Uses

Cooking innovations and the borrowing of traditional recipes are far from being exhausted (Greenberg and Lambert Ortiz 1983, Szita 1987). Old ideas have been retried in seasoning vermouths prepared from the sand pear (*Pyrus pyrifolia*) and in dairy products (Attri *et al.* 1993, Sen and Rajorhia 1994). Saffron dyes are used today for colouring carpets, hats and traditional women's costumes in Sardinia (Campanelli 1990).

Histochemistry

Saffron is used in combination with hematoxylin and phloxine (HPS) to improve animal and human histological staining methods (Du *et al.* 1991, Garvey 1991, Gherardi *et al.* 1987, Levine *et al.* 1988, Martin *et al.* 1992, Sapienza *et al.* 1991, Solsberg *et al.* 1990, van Putten and van Zwieten 1988). Note that wholesale Spanish saffron, 2% extracted in absolute ethanol at 60°C for two weeks, is eight times less expensive than that produced for laboratory dye (Garvey 1991).

Experiments in Medicinal Uses of Saffron

The traditional medicinal uses of saffron have been gradually abandoned. In Italy, for example, the "saffron" entry was first omitted from the "Pharmacopoeia Ufficiale Italiana" in 1964 (Bergaglio 1990). From time to time, however, we witness surges of interest in saffron in normative medical science (Basker and Negbi 1983, Abdullaev 1993).

Folklore relating to the uses of saffron treats its effect on blood clots (Szita 1987). Research at the Institute for Oriental Medicine in Hyogo has shown that saffron stigmas inhibit blood coagulation, via their effect on platelet-aggregation, and accelerate *in vitro* fibrinolysis activity of urokinase and plasmin (Nishio *et al.* 1987, 1993). The toxicity of saffron extract, known from ancient times (Basker and Negbi 1983), has recently been studied in rabbits and dogs (Babaev *et al.* 1990). Possibly related is the effect of saffron's ethanol extract on learning ability in mice (Zhang *et al.* 1994). Saffron chemotherapy has recently been reviewed by Nair *et al.* (1996). Recent progress relating to saffron in biological and medical research is discussed by Abdullaev and Frenkel (this volume).

CONCLUSIONS

The view expressed by Picci (1987), Galigani (1987), Adamo *et al.* (1987) and Dr Byron Skrubis (Thessaloniki, pers. commun.), that mechanical flower harvesting of *C. sativus* is unfeasible, will remain valid until one of several possible developments enables it (Galigani and Garbati Pegna, Grilli Caiola, this volume). A few of these are:

- Advancing the flowering date by enhancing hysteranthous leaf appearance, possibly using precise temperature and water regimes from corm harvesting (Plessner *et al.* 1989).
- More synchronous flowering combined with hysteranthous leaf appearance is also important.
- Other possibilities are increasing the number and weight of stigmas and producing a long flowering shoot, either by breeding (Chichiriccò 1990) or plant-growth regulators. These will give higher yields of saffron in flowers harvested more easily with higher above-ground leaves.

New technologies for in *vitro* production of the spice itself (Barker 1988, Fakhrai and Evans 1990, Himeno and Sano 1987, Himeno *et al.* 1988, Koyama *et al.* 1988, Lu *et al.* 1992, Otsuka *et al.* 1992, Sano and Himeno, 1987, Sarma *et al.* 1990, 1991, Visvanath *et al.* 1990, Han and Zhang 1993, Plessner and Ziv this volume) could lead to *in vitro* saffron which is identical to, or superior to, its natural, high-quality counterpart. Then harvesting would become redundant, but not saffron!

ACKNOWLEDGEMENTS

This study was supported by The Israel Science Foundation (founded by The Israel Academy of Science and Humanities). The help of Dr E. Werker, Jerusalem, in the study of the shoot apex is acknowledged. The literature retrieval carried out in the Faculty of Agriculture Library and the Market Research Unit of the Ministry of Agriculture, and the useful comments given by D. Basker are gratefully acknowledged.

REFERENCES

Abdullaev, F.I. (1993) Biological effects of saffron. *BioFactors*, **4**, 83–86.

Abdullaev, F.I. and Frenkel, G.D. (1999) Saffron in biological and medical research (this volume).

Abulafia, D. (1982) Crocuses and Crusaders: San Giminiano, Pisa and the Kingdom of Jerusalem. In B.Z. Kedar, H.E. Mayer and R.C. Smail (eds.), *Outremer: Studies in the history of the Crusading Kingdom of Jerusalem.* Yad Ben-Zvi Institute, Jerusalem, pp. 227–233.

Adamo, A., Cozzi, M., Galigani, P.F., Vannucci, D. and Vieri, M. (1987) Fabbisogno di manodopera nelle operazioni colturali dello zafferano. In A. Bezzi (1987) pp. 457–460.

Ait-Oubahou, A. and El-Otmani, M. (1999) Saffron cultivation in Morocco (this volume).

Akhund-Zade, I.M. and Muzaferova, R.Sh. (1975) Study of the effectiveness of gamma irradiation of the saffron. *Radiobiologiya*, **15**, 319–322.

Al-Eisawi, D.M. (1986) Studies on the flora of Jordan, 12. Monocotyledons new to Jordan, with notes on some interesting species. *Kew Bulletin*, **41**, 349–357.

Allan, E.J. and Fowler, M.W. (1985) Biologically active plant secondary metabolites. Prospective for the future. *Chemistry and Industry*, **12**, 408–410.

Alōnso, G.L., Varón, R., Gómez, R., Navaro, F. and Salinas, M.R. (1990) Auto-oxidation in saffron at 40°C and 75% relative humidity. *Journal of Food Science*, **55**, 595–596.

Alōnso, G.L., Salinas, M.R., Esteban-Infantes, F.J. and Sanchez-Fernandez, M.A. (1996) Determination of safranal from saffron (*Crocus sativus* L.) by thermal desorption-gas chromatography. *Journal of Agricultural and Food Chemistry*, **44**, 185–188.

Amigues, S. (1988) Le crocus et le safran sur une fresque de Théra. *Revue Archaeologique*, **2**, 227–242.

Anon. Undated. *The saffron crocus*. Saffron Walden Museum Leaflet, no. 13. Saffron Walden Museum, Saffron Walden, England.

Anon. (1989) The field in which the gold of saffron grows–special attention and extra production means. Adulate (A daily newspaper, Teheran) January **14**, p. 5 (Persian).

Anon. United States Department of Agriculture: Foreign Agricultural Service. *US spice trade*. Circular Series, April 1992, 12–13.

Attri, B.L., Lal, B.B. and Joshi, V.K. (1993) Preparation and evaluation of sand pear vermouth. *Journal of Food Science and Technology India*, **30**, 435–437.

Aucante, P. De Grèce en France: A la recherche du safran perdu. *Vivre au Jardin* September 1989, 101–105.

Azizbekova, N.Sh. and Milyaeva, E.L. (1979) Ontogenesis of saffron crocus (*Crocus sativus*) and changes in stem apices. *Soviet Journal of Developmental Biology*, **9**, 266–271.

Azizbekova, N.Sh. and Milyaeva, E.L. (1999) Saffron cultivation in Azerbaijan (this volume).

Azizbekova, N.Sh., Milyaeva, E.L., Lobova, N.V. and Chailakhyan, M.Kh. (1978) Effects of gibberellin and kinetin on formation of flower organs in saffron crocus. *Soviet Plant Physiology*, **25** (3, part 2), 471–476.

Babaev, R.A., Kasumov, F.Yu., Babaev, Kh.F., Akhmedova, G.Sh. and Khasaeva, E.G. (1990) Study of the pharmacological properties of saffron extract. *Izvestiya Akademii Nauk Azerbaidzhana, Seriya Biologicheskikh Nauk*, **0(1)**, 86–93 (Russian).

Bali, A.S. and Sagwal, S.S. (1987) Saffron – a cash crop of Kashmir. *Agricultural Situation in India*, **41**, 965–968.

Barker, S.A. (1988) Biotechnology from European patent applications. *International Industrial Biotechnology*, **8**, 29–34.

Basker, D. (1993) Saffron, the costliest spice: drying and quality, supply and price. *Acta Horticulturae*, **334**, 86–97.

Basker, D. (1999) Saffron chemistry (this volume).

Basker, D. (1999) Saffron technology (this volume).

Basker, D. and Negbi, M. (1983) The uses of saffron. *Economic Botany*, **37**, 228–236.

Basker, D. and Negbi, M. (1985) Crocetin equivalent of saffron extracts: comparison of extraction methods. *Journal of The Association of Public Analysts*, **23**, 65–69.

Behzad, S., Razavi, M. and Mahajeri, M. (1992a) The effect of various amounts of ammonium phosphate and urea on saffron production. *Acta Horticulturae*, **306**, 306–339.

Behzad, S., Razavi, M. and Mahajeri, M. (1992b) The effect of mineral nutrients (N.P.K.) on saffron production. *Acta Horticulturae*, **306**, 426–430.

Bergaglio, G.C. (1990) Note storiche medico-farmaceutiche sullo zafferano. In F. Tammaro and L. Marra (1990) pp. 223–232.

Bezzi, A., ed. (1987) Atti, *Convegno sulla coltivazione delle piante officinali [Convention on the cultivation of medicinal plants]. Proceedings of a conference held in Trento, Italy, 9–10 October* 1986. Istituto Sperimentale per l'Assestamento Forestale e per l'Alpicoltura, Villazzano (Trento).

Brighton, C.A. (1977) Cytology of *Crocus sativus* L. and its allies (Iridaceae). *Plant Systematics and Evolution*, **128**, 137–157.

Campanelli, R. (1990) Storia minima, usi e curiosita dello zafferano in Sardegna. In F. Tammaro and L. Marra (1990) pp. 297–300 (Italian, English abstract)

Chen, L. and Sze, T. (1977) Introduction of *Crocus sativus* into Peking area, China. Observations of its biological characteristics. *Acta Botanica Sinica*, **19**, 313–314.

Chichiriccò, G. (1984) Karyotype and meiotic behaviour of the triploid *Crocus sativus* L. *Caryologia*, **37**, 233–239.

Chichiriccò, G. (1987) Megasporogenesis and development of embryo sac in *Crocus sativus* L. *Caryologia*, **40**, 59–69.

Chichiriccò, G. (1989a) Fertilization of *Crocus sativus* ovules and development of seeds after stigmatic pollination with *C. thomasii* (Iridaceae). *Giornale Botanico Italiano*, **123**, 31–37.

Chichiriccò, G. (1989b) Microsporogenesis and pollen development in *Crocus sativus* L. *Caryologia*, **42**, 249–257.

Chichiriccò, G. (1990) Sterility and improvement of saffron crocus. In F. Tammaro and L. Marra (1990) pp. 99–107.

Chichiriccò, G. and Grilli Caiola, M. (1986) *Crocus sativus* pollen germination and pollen tube growth *in vitro* and after intraspecific and interspecific pollination. *Canadian Journal of Botany*, **64**, 2774–2777.

Chichiriccò, G. and Grilli Caiola, M. (1987) *In vitro* development of parthenocarpic fruits of *Crocus sativus* L. *Plant Cell, Tissue and Organ Culture*, **11**, 75–78.

Chitt, M.A., Gerard, M. and Marechal, J. (1985) La culture des plantes medicinales, aromatiques et condimentaires dans le sud du Maroc: compte-rendue d'un voyage dans les zones d'Agadir, Marrakech, Quarzazate et Tata. *Tropicultura*, **3**, 29–32.

Chrungoo, N.K. and Farooq, S. (1984) Influence of gibberellic acid and naphtaleneacetic acid on yield and on growth in saffron crocus (*C. sativus*). *Indian Journal of Plant Physiology*, **27**, 201–205.

Cichelli, A. (1990) Recenti acquisizioni analitiche nella valutazione dello zafferano. In F. Tammaro and L. Marra (1990) pp. 151–162 (Italian, English abstract.).

Coppock, H.C. (1984) *The saffron crocus in Cherry Hinton and other areas of Cambridgeshire*. D.O. Bond, Cambridge.

Corti, P., Mazzeri, E., Ferri, S., Franchi, G.G. and Dreassi, E. (1996) High performance thin layer chromatographic quantitative analysis of picrocrocin and crocetin, of saffron (*Crocus sativus* L. – Iridaceae): a new method. *Phytochemical Analysis*, **7**, 201–203.

Dafni A., Shmida, A. and Avishai, M. (1981) Leafless autumnal-flowering geophytes in the Mediterranean region–phytogeographical, ecological and evolutionary aspects. *Plant Systematics and Evolution*, **137**, 181–193.

De Mastro, G. and Ruta, C. (1993) Relations between corm size and saffron (*Crocus sativus* L.) flowering. *Acta Horticulturae*, **344**, 512–517.

Dhar, A.K. and Sapru, R. (1993 and 1994) Studies on saffron in Kashmir III. In vitro production of corm and shoot like structures. *Indian Journal of Genetics and Plant Breeding*, **53**, 193–196.

Dhar, A.K., Sapru, R. and Rekha, K. (1988) Studies on saffron in Kashmir I. Variation in natural population and its cytological behaviour. *Crop Improvement*, **15**, 48–52.

Di Crecchio, R. (1960) Lo zafferano. *L'Italia Agricola*, **97**, 629–649.

Di Francesco, L. (1990) *Lo zafferano*. Edizioni L'Informatore Agrario, Verona.

Du, T., Sapienza, S., Eidelman, D.H., Wang, N.S. and Martin, J.G. (1991) Morphometry of the airways during the late responses to antigen challenge in the rat. *American Review of Respiratory Disease*, **143**, 132–137.

Estilai, A. (1978) Variability in saffron (*Crocus sativus* L.). *Experientia*, **34**, 527.

Fakhrai, F. and Evans, P.K. (1990) Morphogenic potential of cultured floral explants of *Crocus sativus* L. for the *in vitro* production of saffron. *Journal of Experimental Botany*, **41**, 47–52.

Ferrazzi, P. (1991) Croco. *L'apicoltore Moderno*, **82**, 75–80.

Fois Sussarello, M.L. (1990) Lo zafferano in Sardegna. In F. Tammaro and L. Marra (1990) pp. 166–170 (Italian, English abstract).

Fritsch, É. (1939) Ethologie du *Crocus* (safran). *Bulletin de la Societe Royale de Botanique de Belgique*, **71**, 104–111.

Fritsch, É. (1942) Ethologie du *Crocus* (safran) (suite). *Bulletin de la Societe de Botanique de Belgique*, **74**, 154–163.

Galigani, P.F. (1987) La meccanizzazione delle colturie di salvia, lavanda, zafferano e genziana. In A. Bezzi (1987) pp. 221–235 (Italian, English summary).

Galigani, P.F. and Garbati Pegna, F. (in press) Mechanized saffron cultivation, including harvesting (this volume).

Garvey, W. (1991) Modification of the Mayer hematoxylin stain. *Journal of Histotechnology*, **14**, 163–165.

George, P.S., Visvanath, S., Ravishankar, G.A. and Venkataraman, L.V. (1992) Tissue culture of saffron (*Crocus sativus* L.): somatic embryogenesis and shoot regeneration. *Food Biotechnology* (New York) **6**, 217–223.

Ghaffari, S.M. (1986) Cytogenetic studies of cultivated *Crocus sativus* (Iridaceae). *Plant Systematics and Evolution*, **153**, 199–204.

Gherardi, G., Del Tacca, M., Paparelli, A., Bernardimi, C. and Polloni, A. (1987) Staining methods for morphometric studies of parietal and gastrin cells in the rat stomach. *International Journal of Tissue Reactions*, **9**, 499–508.

Goliaris, A. (in press) Saffron cultivation in Greece (this volume).

Greenberg, S. and Lambert Ortiz, E. (1983) *The spice of life*. Michael Joseph/Rainbird, London.

Greenberg-Kaslasi, D. (1991) *Vegetative and reproductive development in the saffron crocus (Crocus sativus L.)*. M.Sc. Thesis, The Faculty of Agriculture, The Hebrew University of Jerusalem, Rehovot (Hebrew, English abstract).

Grilli Caiola, M. (1994) Pollen structure and germination of *Crocus thomasii* Ten. (Iridaceae). *Giornale Botanico Italiano*, **128**, 869–877.

Grilli Caiola, M. (1995) A study on pollen grains of *Crocus cartwrightianus* (Iridaceae). *Plant Systematics and Evolution*, **198**, 155–166.

Grilli Caiola, M. (in press) Reproduction biology of saffron and its allies (this volume).

Grilli Caiola, M., Banas, M. and Canini, A. (1993) Ultrastructure and germination percentage of *Crocus biflorus* Miller subsp. *biflorus* (Iridaceae) pollen. *Botanica Acta*, **106**, 488–495.

Halevy, A.H. (1986) The induction of contractile roots in *Gladiolus grandiflorous*. *Planta*, **167**, 94–100.

Halevy, A.H. (1990) Recent advances in control of flowering and growth habit of geophytes. *Acta Horticulturae*, **266**, 35–42.

Han, L.L. and Zhang, X.Y. (1993) Morphogenesis of style-stigma-like structures from the floral explants of *Crocus sativus* and identification of the pigments. *Acta Botanica Sinica*, **35**(supplement), 157–160 (Chinese).

Himeno, H. and Sano, K.T.I. (1987) Synthesis of crocin, picrocrocin and safranal by saffron stigma-like structures proliferated *in vitro*. *Agricultural and Biological Chemistry*, **51**, 2395–2400.

Himeno, H., Matsushima, H. and Sano, K. (1988) Scanning electron microscopic study on the *in vitro* organogenesis of saffron stigma–and style-like structures. *Plant Science*, **58**, 93–101.

Homes, J., Legros, M. and Jaziri, M. (1987) *In vitro* multiplication of *Crocus sativus* L. *Acta Horticulturae*, **212** (II), 675–676.

Iborra, J.L., Castellar, M.R., Cánovas, M. and Manjón, A. (1992) TLC preparative purification of picrocrocin, HTCC and crocin from saffron. *Journal of Food Science*, **57**, 714–716 and 731.

Ingram, J.S. (1969) Saffron (*Crocus sativus* L.). *Tropical Sciences*, **11**, 1771–1784.

Isa, T. and Ogasawara, T. (1988) Efficient regeneration from the callus of saffron (*Crocus sativus*). *Japanese Journal of Breeding*, **38**, 371–374.

Jossen, E. and Stork, A.L. (1983) Oublié, retrouvé: le safran en Valais. *Revue Horticole Suisse*, **56**, 267–274.

Kamikura, M. and Nakazato, K. (1984) Comparison of natural yellow colors extracted from saffron (*Crocus sativus* Linne) and gardenia fruit (*Gardenia jasminoides* Ellis). *Eisei Shikenjo Hokoku* [Bulletin of the National Institute of Hygienic Sciences] **103**, 157–160 (Japanese, English summary).

Kawatani, T., Fujita, S., Kuboki, N. and Saito, K. (1961) Comparison between room and garden cultures of saffron (*Crocus sativus* L.). *Eisei Shikenjo Hokoku* [Bulletin of the National Institute of Hygienic Sciences] **79**, 137–145 (Japanese, English summary).

Kerndorf, H. (1988) Observations on crocus (Iridaceae) in Jordan with special reference to *Crocus moabiticus*. *Herberita*, **44**, 33–53.

Koul, K.K. and Farooq, S. (1984) Growth and differentiation in the shoot apical meristem of saffron plant (*Crocus sativus* L.). *Journal of the Indian Botanical Society*, **63**, 153–169.

Koyama, A., Ohmori, Y., Fujioka, Y., Miyagawa, N., Yamasaki, H. and Kohda, H. (1988) Formation of stigma-like structures and pigment in cultured tissues of *Crocus sativus*. *Planta Medica*, **54**, 375–376.

Laneri, U. (1990) Biotechnological applications to saffron (*Crocus sativus* L.): *in vitro* culture and mutagenesis. In F. Tammaro and L. Marra (eds.), *Lo Zafferano: Proceedings of the International Conference on Saffron (Crocus sativus L.) L'Aquila (Italy) 27–29 October 1989*, Universitá Degli Studi L'Aquila e Accademia Italiana della Cucina, L'Aquila, pp. 109–124.

Levine, D.S., Surawicz, C.M., Ajer, T.N., Dean, P.J. and Rubin, C.E. (1988) Diffuse excess mucosal collagen in rectal biopsies facilitates differential diagnosis of solitary rectal ulcer syndrome from inflammatory bowel diseases. *Digestive Diseases and Sciences*, **33**, 1345–1352.

López Camacho, J.A. and Fucikovsky, Z.L. (1985) Soft rot of the stem of the saffron crocus in the area of Chapingo, Mexico. *Revista Mexicana de Fitopatologia*, **3**, 59 (Spanish).

Lu, X.G., Din, F.M. and Hong, Z.Y. (1988) Preliminary report on introduction of *Crocus sativus* L. *Journal of Zhejiang Forestry Science and Technology*, **8**, 14–18 (Chinese, English summary).

Lu, W.L., Tong, X.R., Zhang, Q. and Gao, W.W. (1992) Study on *in vitro* regeneration of style-stigma like structure in *Crocus sativus* L. *Acta Botanica Sinica*, **34**, 251–256 (Chinese).

Madan, C.L., Kapur and Gupta, B.M. U.S. (1966) Saffron. *Economic Botany*, **20**, 377–385.

Marinatos, S. (1972) *Excavations at Thera V (1971 season)*. Greek Archaeological Society, Athens.

Martin, J.G., Opazo-Saez, A., Du, T., Tepper, R. and Eidelman, D.H. (1992) *In vivo* airway reactivity: predictive value of morphological estimates of airway smooth muscle. *Canadian Journal of Physiological Pharmacology*, **70**, 597–601.

Marzi, V. (1990) Piante aromatiche e medicinali, specie di interesse agrario e possibilità del settore. *Agricoltura delle Venezie*, **44**, 321–331.

Mathew, B. (1977) *Crocus sativus* and its allies (Iridaceae). *Plant Systematics and Evolution*, **128**, 89–103.

Mathew, B. (1982) *The crocuses: a revision of the genus Crocus (Iridaceae)*. B.T. Batsford, London.

Mathew, B. (1984) Crocus L. In P.H. Davis (ed.), *Flora of Turkey and the east Aegean islands*, Vol **8** pp. 413–438, Edinburgh University Press, Edinburgh.

Milyaeva, E.L. and Azizbekova, N.Sh. (1978) Cytophysiological changes in the course of development of stem apices of saffron crocus. *Soviet Plant Physiology*, **25**(2, part 1), 227–233.

Milyaeva, E.L., Azizbekova, N.Sh., Komarova, E.H. and Akhundova, D.D. (1995) *In vitro* development of regenerating corms in *Crocus sativus*. *Fiziologia Rastenii*, **42**, 127–134 (Russian).

Mir, G.M. (1992) *Saffron agronomy in Kashmir: a study in habitat, economy and society*. Gulshan Publishers, Srinagar.

Nair, S.C., Kurumboor, S.K. and Hasegawa, J.H. (1996) Saffron chemoprevention in biology and medicine. A review. *Cancer Biotherapy*, **10**, 257–264.

Nair, S.C., Salomi, M.J., Varghese, C.D., Panikkar, B. and Panikkar, K.R. (1992) Effect of saffron on thymocyte proliferation, intracellular glutathione levels and its antitumor activity. *Biofactors*, **4**, 51–54.

Nauriyal, J.P., Gupta, R. and George, C.K. (1977) Saffron in India. *Arecanut and Spice Bulletin*, **8**, 59–72.

Negbi, M. (1989) Theophrastus on geophytes. *Botanical Journal of the Linnean Society*, **100**, 15–43.

Negbi, M. (1990) Physiological research on the saffron crocus (*Crocus sativus*). In F. Tammaro and L. Marra (1990) pp. 183–207.

Negbi, M., Dagan, B., Dror, A. and Basker, D. (1989) Growth, flowering, vegetative reproduction and dormancy in the saffron crocus (*Crocus sativus* L.). *Israel Journal of Botany*, **38**, 95–113.

Negbi, M. and Negbi, O. (Not dated) The painted plaster floor of Tel Kabri palace: reflections on saffron domestication in the Aegean Bronze Age. Accepted for publication.

Nishio, T., Ogata, K., Okugawa, H., Matsumoto, K., Moriyasu, M. and Kato, A. (1993) Effect of crocus (*Crocus sativus* LINNE, Iridaceae) extract on fibrinolytic enzyme activities (plasmin and urokinase) measured by synthetic substrate method. *Shoyakugaku Zasshi*, **47**, 321–325 (Japanese).

Nishio, T., Okugawa, H., Kato, A., Hashimoto, Y., Matsumoto, K. and Fujioka, A. (1987) Effect of crocus (*Crocus sativus* Linne, Iridaceae) on blood coagulation and fibrinolysis. *Shoyakugaku Zasshi*, **41**, 271–276 (Japanese, English abstract).

Oberdieck, R. (1991) Ein Beitrag zur Kenntnis und Analytik von Safran (*Crocus sativus* L.). *Deutsche Lebensmittll-Rundschau*, **87**, 246–252 (German, English summary).

Otsuka, M., Saimoto, H., Murata, Y. and Kawashima, M. (1992) Method for producing saffron stigma-like tissue and method for producing useful components from saffron stigma-like tissue. United States Patent. 1992, US 5085995, 8 pp., A 28.08.89-US-399037, P. 04.02.92.

Picci, V. (1987) Sintesi sulle esperienze di coltivazione di *Crocus sativus* L. in Italia. In A. Bezzi (1987) pp. 119–157 (Italian, English summary).

Plessner, O., Negbi, M. Ziv, M. and Basker, D. (1989) Effects of temperature on the flowering of the saffron crocus (*Crocus sativus* L.): induction of hysteranthy. *Israel Journal of Botany*, **38**, 1–7.

Plessner, O., Ziv, M., and Negbi, M. (1990) *In vitro* corm production in the saffron crocus (*Crocus sativus* L.). *Plant Cell, Tissue and Organ Culture*, **20**, 89–94.

Raines Ward, D. (1988) Flowers are a mine for a spice more precious than gold. *Smithsonian*, **19**, 104–110.

Rees, A.R. (1988) Saffron–an expensive plant product. *Plantsman*, **9**, 210–217.

Sano, K. and Himeno, H. (1987). In vitro proliferation of saffron (*Crocus sativus* L.) stigma. *Plant Cell, Tissue and Organ Culture*, **11**, 159–166.

Sapienza, S., Du, T., Eiedlman, D.H., Wang, N.S. and Martin, J.G. (1991) Structural changes in the airways of sensitized brown Norway rats after antigen challenge. *American Review of Respiratory Diseases*, **144**, 423–427.

Sarma, K.S., Maesato, K., Hara, T. and Sonoda, Y. (1990) *In vitro* production of stigma-like structures from stigma explants of *Crocus sativus* L. *Journal of Experimental Botany*, **41**, 745–748.

Sarma, K.S., Sharada, K., Maesato, K., Hara, T. and Sonoda, Y. (1991) Chemical and sensory analysis of saffron produced through tissue cultures of *Crocus sativus*. *Plant Cell Tissue and Organ Culture*, **26**, 11–16.

Sen, D.C. and Rajorhia, G.S. (1994) Role of saffron in improving storage quality of sandesh. *Indian Journal of Dairy Science*, **47**, 198–202.

Shmida, A. and Dafni, A. (1989) Blooming strategies, flower size and advertising in the "Lily-group" geophytes in Israel. *Herbertia*, **45**, 111–123.

Siddique, M.A.A., Chrungoo, N.K., Koul, K.K., Farooq, S. and Dahar, U. (1989) Saffron–a valuable crop. *Science Spectrum*, June 1989, 279–281.

Skrubis, B. (1990) The cultivation in Greece of *Crocus sativus* L. In F. Tammaro and L. Marra (1990) pp. 171–182.

Solsberg, M.D., Lemaire, C., Resch, L. and Potts, D.G. (1990) High-resolution MR imaging of the cadaveric human spinal cord: normal anatomy. *American Journal of Neuroloradiology*, **11**, 3–7.

Szita, E. (1986) The saffron harvest is in and the US got it. *The New York Times* 1986 December 10, pp. C1 and C6.

Szita, E. (1987) *Wild about saffron: a contemporary guide to an ancient spice*. Saffron Rose, Daly City, CA.

Tajuddin, Saproo, M.L., Yaseen, M. and Husain, A. (1993) Productivity of rose (*Rosa damascena* Mill) with intercrops under temperate conditions. *Journal of Essential Oil Research*, **5**, 191–198.

Tammaro, F. (1990) *Crocus sativus* L. cv. Piano di Navelli–L'Aquila (zafferano dell'Aquila): ambiente, coltivazione, caratteristiche morfometriche, principi attivi, usi. In F. Tammaro and L. Marra (1990) pp. 47–98 (Italian, English abstract).

Tammaro, F. and Di Francesco, L. (1978) *Lo zafferano dell' Aquila*. Istituto di Tecnica e Propaganda Agraria, Roma.

Tammaro, F. and Marra, L. (eds.) (1990) *Lo Zafferano: Proceedings of the International Conference on Saffron (Crocus sativus L.) L'Aquila (Italy) 27–29 October* 1989. Universitá Degli Studi L'Aquila e Accademia Italiana della Cucina, L'Aquila.

Tuveri, B. (1990) Lo coltivazione dello zafferano in Sardegna. In F. Tammaro and L. Marra (1990) pp. 163–165 (Italian, English abstract).

van Putten, L.J. and van Zwieten, M.J. (1988) Studies on prolactin secreting cells in aging rats of different strains. II. Selected morphological and immunocytochemical features of pituitary tumors correlated with serum prolactin levels. *Mechanisms of Ageing and Development*, **42**, 115–127.

Visvanath, S., Ravishankar, G.A. and Venkataraman, L.V. (1990) Induction of crocin, crocetin, picrocrocin, and safranal synthesis in callus cultures of saffron – *Crocus sativus* L. *Biotechnology and Applied Biochemistry*, **12**, 336–340.

Wallach, B. (1989) Water for Morocco's rivers of palms. *Garden* 1989 May/June, 12–17.

Warburg, E.F. (1957) Crocuses. *Endeavour*, **16**, 209–216.

Wilkins, H.F. (1985) *Crocus vernus*, *Crocus sativus*, in A.H. Halevy (ed.), *CRC Handbook of flowering*, vol. **II** pp. 350–355, CRC Press, Boca Raton, Florida.

Yang, J.X. and S.X. Miao. (1985) Observation on flowering habits of saffron (*Crocus sativus* L.). Acta Agriculturae Universitatis Jilinensis (China) **7**, 25–26 (Chinese).

Xiue, X.H. *et al.* (1986) Effect of increasing yield of gibberellin treated *Crocus sativus*. Bulletin of Chinese Materia Medica **11**, 650–652 (Chinese). (Not seen by the author).

Zanzucchi, C. (1987) La ricerca condotta dal consorzio comunale Parmaensi sullo zafferano (*Crocus sativus* L.). In A. Bezzi (1987) pp. 347–395, (Italian, English summary).

Zhang, Y., Shoyma, Y., Sugiura, H. and Sito, H. (1994) Effects of *Crocus sativus* L. on ethanol-induced impairment of passive avoidance in mice. *Biological and Pharmaceutical Bulletin*, **17**, 217–221.

Ziv, M. and Plessner, O. (1999) Developments in corm and spice production *in vitro* (this volume).

END NOTES

1. The data presented in Table 1.2 of Goliaris' chapter (this volume) covers some of that of Skrubis. However, though it deals with the same region in Greece, they differ greatly.
2. Figures differing from those in Table 1.3 appear elsewhere (Bali and Sagwal 1987, for 1972–73, and Dhar, Sapru and Rekha 1988, for 1982). The latter cite C.K. Atal, who gives a total area of about 5,000 ha and yields higher than 5 lb per acre (=5.7 kg per ha). Mir (1992) presents data on cultivation area and production, in Kashmir, from 1968 until 1983 (when 12,600 ha yielded 145,855 kg, an average of 11.6 kg per ha, more than double the data of Siddique *et al.* 1989, Table 1.3). Note that in Kashmir, saffron is grown partly as an under-crop in rose plantations (Tajuddin *et al.* 1993).
3. On saffron's research and cultivation in Sardinia see the proceedings of an International Seminar on Aromatic and Medicinal Plants held in Cagliary on 1994 and published in *Revista Italiana EPPOS* (1996), **19**. There are five articles concerning saffron cultivation in Sardinia (English abstracts are in *Review of Aromatic and Medicinal Plants* (1997), **7** No. 2).
4. In *Nerine sarniensis*, the hysteranthous habit is also controllable (Halevy 1990).
5. Azizbekova *et al.* (1978), Chrungoo and Farooq (1984) and Xiue *et al.* (1986) reported flowering promotion with exogenous growth regulators. We in Israel failed to repeat these results (Greenberg-Kaslasi 1991, Negbi *et al.* 1989).
6. Having 8 or 16 chromosomes.
7. There Grilli Caiola writes: "Experiments carried out in the field led to the finding that although self-outcross- and unpollinated pistils of saffron did not produce fruits and seeds, a capsule was obtained from a saffron plant grown near *C. cartwrightianus* plants (unpublished data)."
8. AFNOR – l'Association francaise de normalisation.

2. BOTANY, TAXONOMY AND CYTOLOGY OF *C. SATIVUS* L. AND ITS ALLIES

BRIAN MATHEW

90 Foley Road, Claygate, KT10 0NB, UK

ABSTRACT   Saffron is produced from the dried styles of *Crocus sativus* L. (Iridaceae) which is unknown as a wild plant, representing a sterile triploid derived from the naturally occurring diploid *C. cartwrightianus* Herbert. These belong to subgenus *Crocus* series *Crocus* which constitutes 9 species: *C. cartwrightianus* and its derivative *Crocus sativus*, *C. moabiticus*, *C. oreocreticus*, *C. pallasii*, *C. thomasii*, *C. hadriaticus*, *C. asumaniae* and *C. mathewii*. The taxonomy of these species and their infraspecific taxa is presented, together with their distribution, ecology and phenology; full descriptions and chromosome counts are provided and there is a key to their identification.

The genus *Crocus*, a member of the large family Iridaceae, comprises some 85 species having an Old World distribution, primarily in Mediterranean Europe and western Asia. The limits of the entire genus lie within the range longitude 10°W to 80°E, latitude 30°N to 50°N. Phytogeographically, the majority of species occur within the Mediterranean floristic region, extending eastwards into the Irano-Turanian region; both of these areas are characterized by cool to cold winters with autumn–winter–spring precipitation and warm summers with very little rainfall; the latter region experiences much colder winters and generally less rainfall. The genus *Crocus* is well adapted to such conditions, the plants being in active growth from autumn to late spring and surviving the summer drought below ground by means of a compact corm. Many species commence their above-ground growth at the onset of autumn rains and flower almost immediately; some of these produce their leaves and flowers concurrently, or nearly so, while others bloom without leaves and delay their leaf production until the onset of warmer weather, usually in spring. These physiological characteristics, together with cytological information and morphological features of the corm tunics, bracts, bracteols, leaves, flowers and seed, have been used to divide the genus into a hierarchy of subgenera, sections and series (Mathew 1982), and to define the species within those infrageneric groupings. This classification is followed here and is repeated below.

CLASSIFICATION [THE POSITION OF THE "SAFFRON GROUP" IS SHOWN IN BOLD]

1. **Subgenus *Crocus*. Type species: *C. sativus* L.**
   A. **Section *Crocus*. Type species: *C. sativus* L.**
      (a) Series *Verni* Mathew. Type species *C. vernus* Hill.

(b) Series *Scardici* Mathew. Type species *C. scardicus* Kos.
(c) Series *Versicolores* Mathew. Type species *C. versicolor* Ker.-Gawl.
(d) Series *Longiflori* Mathew. Type species *C. longiflorus* Raf.
(e) Series *Kotschyani* Mathew. Type species *C. kotschyanus* Koch
(f) **Series *Crocus*. Type species: *C. sativus* L.**

B. Section *Nudiscapus* Mathew. Type species *C. reticulatus* Stev. ex Adams

(g) Series *Reticulati* Mathew. Type species *C. reticulatus* Stev. ex Adams
(h) Series *Biflori* Mathew. Type species *C. biflorus* Mill.
(i) Series *Orientales* Mathew. Type species *C. korolkowii* Regel ex Maw
(j) Series *Flavi* Mathew. Type species *C. flavus* Weston
(k) Series *Aleppici* Mathew. Type species *C. aleppicus* Baker
(l) Series *Carpetani* Mathew. Type species *C. carpetanus* Boiss. & Reut.
(m) Series *Intertexti* (Maw) Mathew. Type species *C. fleischeri* Gay
(n) Series *Speciosi* Mathew. Type species *C. speciosus* M. Bieb.
(o) Series *Laevigati* Mathew. Type species *C. laevigatus* Bory & Chaub.

2. Subgenus *Crociris* (*Schur*) *Mathew*. Mathew. Type species *C. banaticus* Gay

DEFINITION OF SERIES *CROCUS*

Anthers with extrorse dehiscence **[subgenus *Crocus*]**; scape subtended by a membranous prophyll (enclosed and hidden within the sheathing leaves or cataphylls) **[section *Crocus*]**; corm tunics finely fibrous, usually reticulate; flowers autumnal; leaves rather numerous, usually 5–30, appearing with the flowers or shortly after; bracts flaccid, usually not closely sheathing the perianth-tube, membranous, white or ± transparent with no markings; anthers yellow; style branches 3, usually red and often expanded at the apex, entire or at most fimbriate; seed coats covered with a dense mat of papillae. 2n = 12, 14, 16, 26 **[series *Crocus*]**.

SPECIES COMPRISING SERIES *CROCUS*

1. *C. cartwrightianus* Herbert
2. *C. sativus* Linn.
3. *C. moabiticus* Bornm. & Dinsm.
4. *C. oreocreticus* B.L. Burtt
5. *C. pallasii* Gold.
6. *C. thomasii* Ten.
7. *C. hadriaticus* Herbert
8. *C. asumaniae* B. Mathew & T. Baytop
9. *C. mathewii* Kerndorff & Pasche

IDENTIFICATION KEY TO SPECIES OF *CROCUS*, SERIES *CROCUS*

1. Style branches more than ½ as long (actual measurements) as perianth segments . . 2
   Style branches up to ½ as long (actual measurements) as perianth segments . . . 6
2. Perianth segments 1.4–3.3 cm long; style branches (0.5–)1–2.7 cm long 3
   Perianth segments 3.5–5 cm long; style branches 2.5–3.2 cm long. Triploid, 3n = 24. Cultivated or a relic of cultivation . . . . 2. *C. sativus*
3. Throat glabrous; style divided above or below base of anthers 4
   Throat pubescent at point of insertion of filaments; style divided well below base of anthers, in the throat of the flower 5
4. Flowers white, rarely faintly lilac; style divided well above base of anthers. 2n=26. S. Turkey . . . . 8. *C. asumaniae*
   Flowers mid–lilac to purple (albinos very rare), often with a silver or buff exterior; style divided below base of anthers. 2n=16. Crete . . . . 4. *C. oreocreticus*
5. Leaves (6–)14–24(–30), grey-green; corm with fibrous neck usually 5–8.5 cm long. 2n=14. Jordan . . . . 3. *C. moabiticus*
   Leaves (4–)7–12, green; corm with a fibrous neck usually 2–4 cm long. 2n=16. Greece . . . . 1. *C. cartwrightianus*
6. Flowers white, often stained violet-blue or brown base of segments, inside or out, but occasionally white throughout, rarely tinged pale lilac 7
   Flowers lilac to reddish-purple throughout (albinos very rare) 8
7. Corms tunic parallel-fibrous in lower part, weakly reticulate at apex; style branches 6–10 mm long; centre (throat) of flower not yellow, often with a conspicuous violet-blue zone on the inside. 2n=16. S. Turkey . . . . 9. *C. mathewii*
   Corm tunic reticulate-fibrous ± throughout, style branches 10–19 (–20) mm long; throat of flower usually yellow, occasionally white; if dark-stained, usually confined to the exterior of the flower. 2n=16. Greece . . . . 7. *C. hadriaticus*
8. Throat, and often the filaments, pale yellow. 2n=16. Italy, Dalmatia . . 6. *C. thomasii*
   Throat and filaments white or lilac. Balkans to Iran. 2n=12,14,16 . . . 5. *C. pallasii*

1. C. cartwrightianus Herbert in *Bot. Reg.* 29: Misc.: 82 (1843). Type: Greece, Cyclades, 'ex insula Teno', *Cartwright* (K).

Synonyms:
C. sativus Sibth. & Smith, *Prod. Fl. Graeca* 1: 23 (1806), non Linn.
C. graecus Chapp. in *Bull. Soc. Bot. France* 20: 192 (1873).
C. sativus Linn. var. *cartwrightianus* (Herb.) Maw in *Gard. Chron.* 16: 430 (1881).

Corms 10–15(–20) mm in diameter, depressed-globose, rather flattened at the base; tunics fibrous, the fibres very slender and finely reticulated, extended at the apex of the corm into a neck (2–)2.5–3(–4.5) cm long. Cataphylls 3–5, white, membranous. Leaves (4–)7–12, normally synanthous and equalling the flower at anthesis, spreading, green, 1.5–2.5 mm wide, glabrous or ciliate. Flowers autumnal, fragrant, 1–5, pale to deep lilac-purple or white, strongly veined darker, sometimes stained darker at the base of the segments and on the tube, sometimes pure white with no veining

(albinos are frequent in this species); throat white or lilac, pubescent. Prophyll present. Bract and bracteole present, very unequal, white, membranous with long-tapering, rather flaccid tips. Perianth tube 3–5(–7) cm long; segments subequal, 1.4–3.2 cm long, 0.7–1.2 cm wide, oblanceolate or obovate, obtuse. Filaments 3–7 mm long, white or purplish, glabrous or slightly papillose at the base; anthers 10–15 mm long, yellow. Style divided into 3 red clavate branches, each branch (7–)10–27 mm long, equalling or exceeding the anthers and at least half the length of the perianth segments, arising at a point well below the base of the anthers and usually in the throat of the flower. Capsule ellipsoid, 1.5–2.5 cm long, 0.6–0.7 cm wide, raised on a pedicel to 4 cm long (above ground level) at maturity; seeds reddish-brown, irregularly subglobose, 3–4 mm diameter, the raphe showing as an irregular ridge running the length of the seed and ending in a small, pointed caruncle less than 1 mm long; testa covered with a dense mat of long papillae. 2n = 16.
Phenology: Flowering October–December.
Habitat: Open rocky hillsides, sometimes in short turf or in scrub or sparse pine woods on schist, shale, granite or limestone formations, sea level to 1000 metres.
Distribution: Greece: Attica, Cyclades (recorded from Andros, Giaros, Ios, Kythnos, Mykonos, Naxos, Paros, Serifos, Skiros, Syros, Tenos[1]), Crete.

2. C. sativus Linn., *Species Plantarum*: 36 (1753). Type: Herb. Clifford 18, Crocus (BM).

Synonyms:
C. sativus var. *officinalis* Linn., *Sp. Pl.*, ed. 2,1: 50 (1762).
C. officinalis var. *sativus* Huds., *Fl. Anglica*, ed. 2,1: 13 (1778).
C. autumnalis Smith, *Engl. Bot.* 5: t. 343 (1796).
C. sativus var. *cashmirianus* Royle, *Illustr. Bot. Himal.* 372, t. 90, Fig. 1 (1836).
C. orsinii Parl., *Fl. Ital.* 3: 238 (1858).
C. sativus var. *C. orsinii* (Parl.) Maw in *Gard. Chron.* 16: 430 (1881).

Corms to c. 5 cm in diameter, depressed-globose, flattened at the base; tunics fibrous, the fibres very slender and finely reticulated, extended at the apex of the corm into a neck up to 5 cm long. Cataphylls 3–5, white, membranous. Leaves 5–11, normally synanthous, erect, green, 1.5–2.5 mm wide, glabrous or ciliate. Flowers autumnal, fragrant, 1–4, deep lilac-purple with darker veins and a darker violet stain in the throat; throat white or lilac, pubescent. Prophyll present. Bract and bracteole present, very unequal, white, membranous with long-tapering, rather flaccid tips. Perianth tube 4–5(–8) cm long; segments subequal, 3.5–5 cm long, 1–2 cm wide, oblanceolate or obovate, obtuse. Filaments 7–11 mm long, purplish, glabrous; anthers 15–20 mm long, yellow. Style divided into 3 deep red clavate branches, each branch 25–32 mm long, much exceeding the anthers and at least half the length of the perianth segments, arising at a point well below the base of the anthers in the

---

[1] According to Prof. C. Doumas (Athens) and Dr A. Sarpaki (Chania, Crete) *C. cartwrightianus* grows on a volcanic ash near Akrotiri, Santorini, and is being harvested for local consumption by the villagers (editor's note).

throat of the flower. Capsules and seeds rarely produced (a triploid of low fertility). 3n = 24.
Phenology: Flowering October–November.
Distribution: Known only as a cultivated plant, probably of very ancient origin.

Notes: There is a little doubt that this 'species' is derived from, and probably a clonal selection of, *C. cartwrightianus*. In this case, under the International Code of Nomenclature, the name of the latter should change to the earlier-published *C. sativus*. However, for practical purposes it makes good sense to consider the widely cultivated clonal selection as a 'neo-species' and retain the familiar and much-used name *C. sativus* for this plant, which is commercially cultivated as the source of Saffron. This treatment would not be creating a precedent; similar nomenclature has been adopted for other major crop plants which now differ markedly from their wild ancestors, for example Garlic (*Allium sativum*), Onion (*Allium cepa*) and several cereals.

3. C. moabiticus Bornmüller & Dinsmore ex Bornmüller in *Feddes Repert.* 10: 383 (1912). Type: Jordan. Moab, near Zizeh, 720 metres, 18 November 1910, *Meyers & Dinsmore* M. 1537 (B).

Corms 20–35 mm diameter, subglobose, flattened at the base; tunics finely fibrose, the fibres are parallel at the base and weakly reticulate at the apex, extended into a distinct neck (4–)5.5–8.5(–9.5) cm long. Cataphylls 3, white, membranous. Leaves (6–)14–24(–30), usually present but short at flowering time, grey-green, 1–1.5 mm wide, sparsely papillose on the margin of the keel. Flowers 1–6, autumnal, fragrant, veined purple to varying degrees on all six segments on a white ground colour, sometimes so heavily as to appear purple, sometimes stained darker at the base of the segments and on the tube; throat white or purple, pubescent. Prophyll present. Bract and bracteole present, unequal, the bracteole narrower and slightly shorter than the bract, white, membranous with long-tapering, rather flaccid tips. Perianth tube 2–5 cm long, white or purple; segments subequal, 1.5–3.2 cm long, 0.3–1.2 cm wide, narrowly elliptic to oblanceolate or obovate, acute to obtuse. Filaments 2.5 mm long, white ageing to purple, glabrous; anthers 10–15 mm long, yellow. Style divided into 3 deep red clavate branches, branch 15–20 mm long, equalling to much exceeding the anthers and at least half the length of the perianth segments, arising at a point well below the base of the anthers in the throat of the flower. Capsule ellipsoid, 1.5–2.5 cm long, 0.5–0.7 cm wide, carried on a very short pedicel at maturity, sometimes not exceeding the ground level; seeds dark brown, irregularly subglobose, 3–3.5 mm diameter, covered with a dense mat of long papillae. 2n = 14.
Phenology: Flowering November–December.
Habitat: Open rocky hillside on limestone formation in scrub, sparse grass and maquis, 680–950 metres.
Distribution: Jordan, Moab.

4. C. oreocreticus B.L. Burtt in *Phyton* 1: 224 (1949). Type: Crete, 'Hierapetra: Aphendi Kavusi, oberhalb Thriphti, 900–1350 m, 3 December 1939', *Davis* 1609 (K).

Corms ovoid, c. 10–15 mm diameter, depressed-globose and flattened at the base; tunics fibrous, the fibres finely reticulated. Cataphylls 3–4, white or pinkish-brown

stained, membranous. Leaves 7–15, subhysteranthous or synanthous but if absent at anthesis then developing immediately after flowering, green or slightly greyish, 0.5–1 mm wide, glabrous. Flowers autumnal, 1–2, rarely more, mid-lilac to purple with darker veining, the exterior pale silvery or buff coloured (very rarely albino), throat lilac, glabrous. Prophyll present. Bract and bracteole present, subequal in length but the bracteole narrower, white and somewhat flaccid, tapering gradually to an acute apex. Perianth tube usually 4–5 cm long, white or lilac; segments subequal, 1.4–3.3 cm long, 0.4–1.1 cm wide, oblanceolate, obtuse, the inner usually slightly smaller than the outer. Filaments 5–8 mm long, white, glabrous; anthers 10–17 mm long, yellow. Style divided into 3 red (rarely yellow), apically thickened branches, each branch (5–)13–20(21) mm long and about equalling the tips of the anthers, arising at a point at or just above the throat of the flower, below the base of the anthers. Capsule oblong, c. 15 mm long, 7 mm wide, produced on a short pedicel just above ground level; seeds reddish-purple, subglobose, c. 3–4 mm long, with a pointed caruncle; raphe, a rather low ridge running the length of the seed; testa covered with a dense mat of papillae. $2n = 16$.
Phenology: Flowering October–December.
Habitat: Open rocky mountains with *Astragalus, Phlomis, Sarcopoterium spinosum* and *Berberis cretica* in heavy reddish soil on limestone formations, 900–2000 metres.
Distribution: Crete, recorded on Mt. Psiloritis (Ida, Idi), the Lasithi and Katharo areas, and the Sitia mountains.

5. C. pallasii Gold. in *Mém. Soc. Nat. Moscou* 5: 157 (1817).
A very widespread and variable species; four subspecies are recognised as follows:

1. Perianth segments ligulate or narrowly oblanceolate, usually 4–7 mm wide, deep reddish-purple; style branches usually pale yellow.
   Southern Turkey, northern Syria . . . . **5d**. subsp. **dispathaceus**
   Perianth segments obovate or oblanceolate, (4–)8–16 mm wide, pale to deep lilac-blue; style branches usually red or orange 2
2. Style branches widely and abruptly expanded at apex; perianth segments obovate, rounded or obtuse, often notched at apex; corm with a long fibrous neck to 10 cm long.
   Western Iran, north-eastern Iraq, southern Jordan . . . **5c**. subsp. **haussknechtii**
   Style branches expanding gradually to the apex and usually fairly slender throughout; perianth segments elliptic, oblanceolate or obovate, acute to acuminate; corm without a neck or with a fibrous neck up to 6 cm long 3
3. Perianth segments narrowly oblanceolate, acute to acuminate, 4–10(–12) mm wide; corm with an extended fibrous neck (2–)3.5–6 cm long. South-eastern Turkey, Syria, Lebanon . . . . **5b**. subsp. **turcicus**
   Perianth segments elliptic, oblanceolate or obovate, subacute to acute, (5–)8–16 mm wide; corm without a neck or with a short fibrous neck up to 2 cm long . 4
4. Leaves green; corm without an obvious neck. Southern Peloponnese . . . .
   . . . . **5e**. subsp. 'E'
   Leaves grey-green; corm with a neck 1–2 cm long. Balkans, Crimea, western, central and southern Turkey, Syria, Lebanon, Israel . . . . . **5a**. subsp. **pallasii**

5a) C. pallasii subsp. **pallasii** Goldb. in *Mém. Soc. Nat. Moscou* 5: 157 (1817). Type: Crimea, 'in Tauriae collibus' (no specimen traced).

Synonyms:
C. serotinus Ker-Gawl. in Bot. Mag.: t. 1267 (1810), partly as to syn. *C. autumnalis campestris* specimens from Crimea and southern Russia.
C. campestris Pallas ex Herbert in Bot. Mag. sub. t. 3864 (1841).
C. hybernus Friv. in Griseb., Spicil. Fl. Rumel. 2: 374 (1844).
C. pallasianus Herb. in Bot. Reg. 30, t. 3, Fig. 2 (1844).
C. sativus var. *elwesii* Maw in Gard. Chron. n.s. 16: 430 (1881).
C. olbanus Siehe in Allg. Bot. Zeitschr. 12: 1 (1906).
C. elwesii (Maw) O. Schwarz in Fedde Repert. Sp. Nov. 36: 74 (1934).
C. thiebautii Mouterde in Bull. Soc. Bot. France 101: 422 (1954).
C. libanoticus Mouterde in Bull. Soc. Bot. France 101: 422 (1954).
C. haussknechtii sensu Mouterde, Nouv. Fl. Lib. et Syrie 1: 297 (1966), non Boiss. & Reut. ex Maw) Boiss.

Corms ovoid, c. 10–20(–25) mm diameter, depressed-globose; tunics fibrous, the fibres finely reticulated, extended at the apex into a neck up to 2 cm long. Cataphylls 3–5, white, membranous. Leaves (5–)7–17, synanthous or subsynanthous but if absent at anthesis then developing immediately after flowering, greyish-green, 0.5–1.5 mm wide, glabrous or scabrid to papillose on the margins of keel and blade. Flowers fragrant, autumnal, 1–6, pale pinkish-lilac to deep lilac-blue or purplish-blue, usually slightly veined darker; throat white or lilac, pubescent. Prophyll present. Bract and bracteole present, unequal, membranous, white, tapering gradually to acute, flaccid tips. Perianth tube 4–7(–10) cm long, white, lilac or purplish; segments (1.9–)2.5–5 cm long, (5–8)–16 cm wide, elliptic, oblanceolate or obovate, acute or subacute, the inner often slightly smaller than the outer. Filaments 2–5 mm long, white, glabrous or sparsely papillose-pubescent; anthers 9–20 mm long, yellow. Style divided into 3 red (occasionally yellow) branches, each branch 3–15 mm long and half as long as the perianth segments, rather slender and tapering gradually to the expanded apex. Capsule ellipsoid, 15–25 mm long, 7–10 mm wide, produced on a short pedicel at or just above ground level at maturity; seeds reddish-purple, irregularly subglobose, 3–4 mm diameter, with a small, pointed caruncle; raphe, usually a small ridge running the length of the seed but occasionally wing-like; testa covered with a dense mat of papillae. $2n = 14$.
Phenology: Flowering October–November.
Habitat: Open stony or rocky hillsides, often on sparse scrub or spiny steppe vegetation, on limestone or basalt formation, 70–2820 metres.
Distribution: Macedonia, southern Serbia, eastern Roumania, southern and eastern Bulgaria, northern Greece, Aegean Islands (Lesbos), Crimea, Lebanon, Israel, western, central and southern Turkey, ? Syria.

b) C. pallasii subsp. **turcicus** B. Mathew in *Pl. Syst. Evol.* 129: 98 (1977). Type: Turkey, Gaziantep Province, Urfa to Gaziantep road, 22 km from Gaziantep, steep chalky hillside, 800 m, 3 November 1973, *T. Baytop & B. Mathew* ISTE 27025 (K holotype, ISTE isotype).

Synonym:
C. macrobolbos Jovet & Gomb. in *Bull. Soc. Bot. France* 103, 7–8: 460 (1956).
Description as for subsp. *pallasii* except for the following:
Corm 15–35 mm diameter; tunics extended into a neck (2–)3.5–6 cm long. Leaves absent at flowering time but the dried remains of the previous season's sometimes persisting until anthesis. Perianth segments 2.5–5 cm long, 4–10(–12) mm wide, narrowly oblanceolate, acute to acuminate. Filaments 2–4 mm long; anthers 10–12 mm long. 2n = 12.
Phenology: Flowering October–November.
Habitat: Dry regions, usually in rocky places with steppe vegetation, 600–1700 metres.
Distribution: South-eastern Turkey.

c) C. pallasii subsp. **haussknechtii** (Boiss. & Reut. ex Maw) B. Mathew in *Pl. Syst. Evol.* 128: 99 (1977). Type: Iran, 'Kurdistan'.

Synonyms:
C. sativus var. *haussknechtii* Boiss. & Reut. ex Maw in *Gard. Chron.* 16: 430 (1881).
C. haussknechtii (Boiss. & Reut. ex Maw) Boiss., *Fl. Orient.* 5: 100 (1882)

Description as for subsp. *pallasii* except for the following:
Corms up to 30 mm diameter with a fibrous neck up to 10 cm long. Perianth segments obovate, rounded or obtuse, often emarginate or retuse, rarely acute, 3.5–4.2 cm long, 0.8–1.4 cm wide. Filaments 3–6 mm long; anthers 1.3–2 cm long. Style branches intense dark red, 5–13 mm long, clavate, markedly and abruptly expanded at the apex, the point of division of the style varying from a point level with the middle of the anthers to just below their tips. 2n = 16.
Phenology: Flowering October–November.
Habitat: Dry fields or rocky hillside, or in sparse *Quercus* scrub, 1300–2100 metres.
Distribution: Western Iran, north-eastern Iraq, southern Jordan.

d) C. pallasii subsp. **dispathaceus** (Bowles) B. Mathew, *The Crocus*: 54 (1982). Type: ?N. Syria, Aleppo, 17 December 1912. Cultivated specimens from corms distributed by George Egger of Jaffa, Syria.

Synonym:
C. dispathaceus Bowles, *A Handbook of Crocus and Colchicum*, ed. 1: 68 (1924).

Description as for subsp. *pallasii* except for the following:
Corms up to 30 mm diameter with fibrous neck (2–)3–7 cm long. Flowers deep reddish-purple or mauve-pink. Perianth segments 4–7 mm wide, ligulate or very narrowly oblanceolate. Style branches inconspicuous, very slender, yellow or sometimes pale orange. 2n = 14.
Phenology: Flowering September–November.
Habitat: Dry *Quercus coccifera* scrub or in sparse *Juniperus/Quercus/Pinus* woods, in terra rossa on limestone formation, 350–2000 metres.
Distribution: Southern Turkey, northern Syria.

e) C. pallasii subsp. **E** (B. Mathew, ined.). Based on a specimen collected in Greece, Peloponnese, Mt. Parnon, *M. Koenen* s.n. (K).

Description as for subsp. *pallasii* except for the following:
Corms almost without a fibrous neck, the tunics weakly reticulate. Leaves 5–10, green. Flowers bright lilac; style branches orange, arising at a point near or above the top of the anthers. 2n = ??.
Phenology: Flowering ?October–November.
Habitat: Open rock scrubland, 1500 metres.
Distribution: Southern Greece, Peloponnese, Mt. Parnon near Ag. Vasileios, 1550 metres.
Note: Field studies are required; at first this appeared to be close to *C. pallasii* and was tentatively placed here as a further subspecies. However, recent studies now suggest that it should be regarded as a subspecies of *C. hadriaticus*.

6. C. thomasii Ten., *Mem. Crochi Fl. Nap.* 12, t. 4 (1826). Type: Italy. Calabria, Serre di S. Bruno; Monte della Stella, *Thomas* (K).

Synonyms:
C. thomasianus Herb. in *Bot. Reg.* 30: t. 3, Fig. 6 (1844).
C. visianicus Herbert in *Bot. Reg.* 1845: Misc. 83 (1845).

Corms 8–12(–15) mm diameter, depressed-globose, flattened at the base; tunics fibrous, the fibres very slender and finely reticulated, extended at the apex of the corm into a neck up to 1 cm long. Cataphylls 3–5, papery, white. Leaves 5–10, synanthous, usually equalling the flower at anthesis, but sometimes only the tips showing, green, 0.5–1.5 mm wide, glabrous or papillose on the margins. Flowers autumnal, fragrant, 1–2(–3), pale to deep lilac, generally not strongly veined darker but sometimes veined or stained violet towards the base of the segments; throat pale yellow, pubescent. Prophyll present. Bract and bracteole present, very unequal, white, membranous with long tapering flaccid tips. Perianth tube 3–6(–8) cm long; segments 2–4.5 cm long, 0.7–1.5 cm wide, elliptic, obovate or oblanceolate, acute or obtuse. Filaments 5–8 cm long, usually pale yellow, glabrous or finely pubescent at the base; anthers 9–13 mm long, yellow. Style divided at a variable point, usually ranging from just below or level with the base of the anthers to about a quarter of the way up the anthers, into 3 bright red branches, each 0.7–2 cm long, half or less than half the length of the perianth segments, expanded gradually to the apex. Capsule ellipsoid, 1–1.5 cm long, 0.5–0.7 cm wide, raised on a pedicel to 2.5 cm long (above ground level) at maturity; seed globose, about 2 mm diameter with a poorly developed raphe and pointed caruncle. 2n = 16.
Phenology: Flowering: October–November.
Habitat: Open rocky or stony slopes or in sparse scrub, sea level to 1000 metres.
Distribution: Southern Italy and the Dalmatian coastal region.

7. C. hadriaticus Herbert in *Bot. Reg.* 31: Misc. 82 (1845). Type: Greece, Levkas Is., 'On the hill of Chrysobeloni', *Vrioni* (K); Dodona, near Ioannina, *Saunders* (not traced).

Synonyms:
C. hadriaticus var. *chrysobelonicus* Herb., loc. cit.
C. hadriaticus var. *saunderianus* Herb., loc. cit.

C. peloponnensiacus Orph. in Boiss., Diagn. Ser. 2,4: 95 (1895).
C. nivalis Klatt in Linnaea 34: 720 (1865–1866), partly as to syn. *C. peloponnensiacus* Orph. and specimens *Orphanides* 67, 68.

Corms 10–15 mm diameter, depressed-globose, rather flattened at the base; tunics fibrous, the fibres very slender and finely reticulated, extended at the apex of the corm into a short neck. Cataphylls 3–4, white, membranous. Leaves 5–9, normally synanthous, sometimes equalling the flower at anthesis, but sometimes very short and occasionally absent, but then appearing immediately after the flowers, grey-green, 0.5–1 mm wide, ciliate. Flowers autumnal, fragrant, 1–3, white, often stained externally brownish, yellowish or violet at the base of the segments, rarely flushed throughout pale lilac; throat yellow or rarely white, pubescent. Prophyll present. Bract and bracteole present, subequal or with the bracteole much narrower, white, membranous with long-tapering, rather flaccid tips. Perianth tube 3–9 cm long, white, yellow, brownish or violet; segments equal or the inner slightly smaller, 2–4.5 cm long, 0.7–2 cm wide, elliptic-oblanceolate, obtuse. Filaments 3–11 cm long, yellow or white, glabrous or sparsely and minutely pubescent just at the base; anthers 7–15 mm long, yellow. Style divided into 3 slender branches, each branch 10–16(–20) mm long, slightly shorter than or exceeding the anthers, less than half the length of the perianth segments, arising at a point above the throat of the flower. Capsule ellipsoid, 1.2–2 cm long, 0.6–0.8 cm wide, raised on a pedicel to 4.5 cm long (above ground level) at maturity; seed reddish-brown, subglobose, 2–3 mm long, the raphe narrow and poorly developed, caruncle pointed, less than 1 mm long; testa covered with a dense mat of papillae. $2n = 16$.
Phenology: Flowering: September–November.
Habitat: In open scrub or short turf or rock hillsides of limestone or shale, 250–1500 metres.
Distribution: Western and southern Greece, recorded from the Pindus Mountains, Mt. Parnassus, central, southern and eastern Peloponnese, Cephalonia Is., Levkas Is., Kythira Is.
Notes: The pale lavender-coloured variants of *C. hadriaticus* from the southern Peloponnese have been named forma *lilacinus* B. Mathew in Kew Mag. 3,4: 311 (1986). These occur sporadically in populations of otherwise 'normal' white-flowered plants and may be the result of introgression from another taxon in the area, possibility that referred to above as *C. pallasii* subsp. 'E'. Recent studies now suggest that this represents a subspecies of *C. hadriaticus* B. Mathew, ined. The wholly white-flowered plants (i.e. lacking a yellow throat) from Mt. Parnassus are named forma *parnassicus* B. Mathew, loc. cit.

8. C. asumaniae B. Mathew & T. Baytop in *Notes Roy. Bot. Gard. Edinb.* 37, 3: 469 (1979). Type: Turkey, Antalya Province, near Aseki, 900 m, 6 November 1976, T. *Baytop* ISTE 36254 (K holotype, E, ISTE isotypes).

Corms ovoid, c. 15–20 mm diameter; tunics fibrous, the fibres very slender and finely reticulated, extended at the apes of the corm into a neck 3–4 cm long. Cataphylls 2–3, white, membranous. Leaves 5–6, hysteranthous or with the tips just showing at anthesis, slightly greyish-green, 0.5–1 mm wide, glabrous. Flowers autumnal, 1–3,

white, occasionally with dark veins near the base of the segments, rarely very pale lilac; throat whitish or pale yellow, glabrous. Prophyll (?absent acc. to *The Crocus*) present. Bract and bracteole present, unequal, white, membranous with long-tapering, rather flaccid tips. Perianth tube 5–8 cm long, white; segments subequal, 2.5–3 cm long, 0.5–1 cm wide, oblanceolate or narrowly elliptic, obtuse to acute, the inner slightly smaller than the outer. Filaments 2–5 mm long, white or pale yellow, glabrous; anthers 10–20 mm long, yellow. Style divided into a reddish-orange clavate branches, each branch 13–20 mm long and considerably exceeding the anthers and at least half the length of the perianth segments, arising at a point well above the base of the anthers. Capsule ellipsoid, c. 1 cm long; seeds reddish-purple, subglobose, 2–3 mm long, with a pointed caruncle about 1 mm long; raphe, a rather indistinct ridge running the length of the seed; testa covered with a dense mat of long papillae. $2n = 26$.
Phenology: Flowering October–November.
Habitat: Open spaces in *Quercus cerris* and *Q. coccifera* scrub, in stony ground with limestone outcrops, 900–1250 metres.
Distribution: Turkey, Antalya Province.

9. C. mathewii Kerndorff & Pasche in *The New Plantsman* 1,2: 102–106 (1994). Type: Turkey, Antalya Province, Lycian Taurus Mts., 400–1100 metres, 16 November 1992, *Kerndorff & Pasche* HKEP 9291 (holotype K).

Corms (13–)16(–24) mm diameter, depressed-globose, flattened at the base; tunics fibrous, the fibres slender and parallel in the lower part, slightly reticulate near the apex of the corm, extended into a neck (10–)19(–32) cm long. Cataphylls 2–4, silvery-white, membranous, suffused brown near the apex. Leaves (4)7(10), hysteranthous, dark green, slightly greyish, 1–2 mm wide, sparsely ciliate. Flowers autumnal, fragrant, 1–3, white or rarely pale lilac-blue, often stained deep violet at the base of the segments inside and outside; throat violet, pubescent. Prophyll present. Bract and bracteole present, subequal, silvery-white, membranous with long-tapering, rather flaccid tips. Perianth tube (4–)7(–12) cm long, usually violet in the upper part, paler to almost white lower down; segments subequal, 1.9–3 cm long, 0.7–1.3 cm wide, ovate to obovate, obtuse to slightly acuminate, the inner slightly smaller than the outer. Filaments 3–4 mm long, white, glabrous; anthers 10–12 mm long, yellow. Style divided into 3 orange to red branches, each branch 6–10 mm long, usually clearly exceeding, but sometimes equalling or rarely shorter than, the anthers, and less than half as long (rarely half as long) as the length of the perianth segments, arising at a point well above the base of the anthers. Capsule ellipsoid, c. 2 cm long and 1 cm wide, raised on a short pedicel above ground level at maturity; seeds purplish-brown, globose, 4–5 mm diameter, the raphe an indistinct ridge, caruncle pointed, less than 1 mm long; testa covered with a dense mat of papillae. $2n = 16$ (M. Johnson, pers. comm.)
Phenology: Flowering October–November.
Habitat: In *Quercus coccifera* scrub, between dolomite and calcareous rocks, 400–1000 metres.
Distribution: Turkey, Antalya and Muğla Provinces.

Notes: In Antalya and Muğla Provinces there are populations of *Crocus* of this alliance which require detailed field investigation; the flowers are white or very pale lilac, often without the very conspicuous violet zoning which is such a striking feature of 'typical' *C. mathewii*. It is possible that these represent populations of *C. mathewii* in which there has been some introgression from another species, perhaps *C. pallasii*.

ACKNOWLEDGMENTS

I am indebted to the many friends and colleagues who have given me information on *Crocus* species in the wild, and living material for study. In connection with the 'Saffron group' I must mention in particular Turhan Baytop, Peter Bird, Chris Brickell, Ray Cobb, Erna Frank, Chris Lovell, Helmut Kerndorff, Manfred Koenen, Erich Pasche, Jimmy Persson, Mike Salmon, David Stephens, Bob and Rannveig Wallis and Martin Young. I would also like to thank Christine Heywood (then Christine Brighton) for the extensive cytological investigations carried out on *Crocus* whilst working in the Jodrell Laboratory at Kew; also Margaret Johnson for carrying out chromosome studies of the taxa described more recently.

REFERENCES AND BIBLIOGRAPHY

Bowles, E.A. (1924) *A Handbook of Crocus and Colchicum*. The Bodley Head, London.

Brighton, C.A. (1977) Cytology of *Crocus sativus* and its allies. *Plant Systematics and Evolution*, **128**, 137–157.

Burtt, R.L. (1948) *Crocus oreocreticus*. *Phyton*, **1**, 224–225.

Feinbrun, N. (1957) The genus *Crocus* in Israel and neighbouring countries. *Kew Bulletin*, **12**, 270–276.

Feinbrun, N. and Shmida, A. (1997) A new review of the genus *Crocus* in Israel and neighbouring countries. *Israel Journal of Botany*, **26**, 172–189.

Herbert, W. (1847) History of the species of *Crocus*. *Journal of the Horticultural Society of London*, **2**, 249–293.

Kerndorff, H. (1988) Observations on *Crocus* (Iridaceae) in Jordan with special reference to *Crocus moabiticus*. *Herberita*, **44**, 33–53.

Mathew, B. (1982) *The Crocus — A Revision of the Genus Crocus*. Batsford, London.

Mathew, B. and Baytop, T. (1976) Some observations on Turkish *Crocus*. *Notes from the Royal Botanic Gardens, Edinburgh*, **35**, 61–67.

Maw, G. (1886) *A Monograph of the Genus Crocus*. Dulau and Co., London.

Mouterde, P. (1966) *Nouvelle flore du Liban et de la Syrie*. Imprimerie Catholique, Beyrouth, Vol. **1**, 295–299.

3. REPRODUCTION BIOLOGY OF SAFFRON AND ITS ALLIES

MARIA GRILLI CAIOLA

Department of Biology, University of Rome "Tor Vergata,"
Via della Ricerca Scientifica 1 — 00133 Rome, Italy

ABSTRACT   Studies on the reproductive biology of saffron and its allies (*C. thomasii* and *C. cartwrightianus*) have indicated that the triploid *C. sativus* is mainly male sterile whereas the diploid *C. thomasii* and *C. cartwrightianus* are self-sterile but cross-fertile. Saffron pollen is anomalous and a high percentage is unviable, and *in vitro* and *in vivo* it has a very low percentage of germinating grains. Experiments have demonstrated that its pistil can be fertilized by the pollen of *C. thomasii* or *C. cartwrightianus*. Calcium ion seems to be involved in pollen-tube growth and the fertilization process in fertile *Crocus* species.

INTRODUCTION

Saffron (*Crocus sativus* L., Iridaceae) multiplies by means of corms, and man selects the best ones for cultivation. Because saffron occurs only in culture, corm selection has led to improved populations of some characteristics, mainly the long red stigmas which, once dried, form the commercially important spice.

Vegetative multiplication offers advantages in maintaining the genetic characteristics of the plant, but it does not allow for any genetic improvement. Thus saffron from different cultivation areas represents clones which differ only in minor morphological and biochemical characteristics. Recent investigations by flow cytometry on the DNA of saffron cultivated in Italy, Israel and Spain (Brandizzi and Grilli Caiola 1996a) demonstrated their DNA to be very similar quantitatively and in their qualitative base composition, despite the fact that saffron from Israel was quite different from the others in flower morphology. Commercial saffron is obtained from dried stigmas in which secondary products such as the crocin and crocetin carotenoids, and the bitter principles derived from safranal, are concentrated. Saffron production is a very long and expensive process due to the reduced number of flowers formed on each corm. Some attempts have been made to obtain corms and flowers *in vitro* (Igarashi *et al.* 1993), but for genetic improvement, seeds are needed. To date, seed set from saffron have not been reliably reported due to the plant's triploid genome. Studies on saffron reproduction were scarce until recent years, when many studies were initiated with the aim of comparing the infertile saffron with its supposed fertile diploid ancestors.

Little is known about the possible ancestors of saffron and how saffron originated. Comparative morphological, cytological and phenological studies (Brighton 1977, Mathew 1982, Karasawa 1933) led to the hypothesis that the most probable ancestors of *C. sativus* (Figure 3.1) were *C. cartwrightianus* Herb. (Figure 3.2) or *C. thomasii* Ten. (Figure 3.3). These last two species have different natural distribution areas and the likelihood of natural crosses between saffron and these species is improbable. The hypothesis that *C. sativus* is related to *C. cartwrightianus* has been confirmed by recent data obtained by flow cytometry, quantitative and qualitative analysis of DNA of isolated nuclei, and mathematical DNA-base-pair estimation (Brandizzi and Grilli Caiola 1996a). The DNA content of *C. cartwrightianus* nuclei is very similar to *C. sativus*, less so with respect to *C. thomasii*. Although analysis of G-C and A-T content also indicates higher G-C content in all three species, the G-C percentage in *C. sativus* was homogeneous in different clones and similar to *C. cartwrightianus*, but lower than that in *C. thomasii*. Polisomaty did not occur in any of the three species.

However, taking into account that on an embryological basis (Chichiriccò 1989a) *C. thomasii* could also be an ancestor of saffron, attempts were made to compare reproduction processes in the three species. *C. sativus*, *C. cartwrightianus*, *C. thomasii* were compared with respect to male and female gametophytes, and pollen-tube development in the stigma and style after self-, cross-, intra- and interspecific pollination (Grilli Caiola *et al.* 1985, Grilli Caiola 1994, 1995). Preliminary studies considered the reproduction biology of *C. sativus*, *C. cartwrightianus* and *C. thomasii*, then experiments were carried out on mixed cultures of *C. sativus* with *C. thomasii* and *C. sativus* with *C. cartwrightianus*.

Figure 3.1 *Crocus sativus.*

Figure 3.2 *Crocus cartwrightianus*, white-flowered.

Figure 3.3 *C. thomasii*.

SPOROGENESIS

Saffron triploidy has been demonstrated by numerous authors (e.g., Karasawa 1933, Brighton 1977). Studies on micro- and megasporogenesis in saffron confirmed that meiosis occurs in an anomalous manner with irregular chromosome pairing, division and distribution in the derived nuclei (Chichiriccò and Grilli Caiola 1984, Chichiriccò 1987, 1989b, Grilli Caiola and Chichiriccò 1991). Yet whereas microsporogenesis gives rise to a high percentage of anomalous microspores and consequently different pollen grains, megaspores and the derived embryo sac developed in a more regular manner. Thus a high number of regular embryo sacs have been found in the ovules of saffron. Regular micro- and megasporogenesis occur in the diploid *C. thomasii* and *C. cartwrightianus*, and a high percentage of the spores and derived gametophytes are therefore normally structured.

POLLEN ORGANIZATION AND VIABILITY

C. sativus mature pollen grains (Figure 3.4) are roundish in shape, with a few being ovoid. Their dimensions vary: among grains of 91–94 µm in diameter, some are small, 50 µm in diameter. Pollen grains are inaperturate with a finely multiaperturate exine, covered by numerous spinulae and lipid bodies forming pollenkitt (Figure 3.5). Pore-like depressions without exine but surrounded by more numerous spinulae and lipid droplets are randomly distributed. In many cases, the pollen grain wall shows very large bands of broken exine. The median section of a pollen grain observed

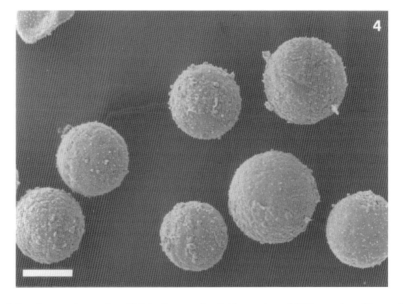

Figure 3.4 *Crocus sativus* pollen, SEM micrographs, Pollen grains of different sizes [Bar = 50 µm].

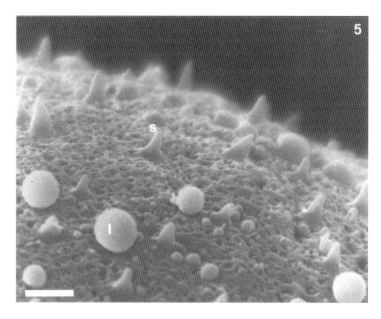

Figure 3.5 *Crocus sativus* pollen, SEM micrographs, Exine surface with spinulae (s) and lipid droplets (I) [Bar = 2 μm].

with a transmission electron microscope (TEM, Figure 3.6) shows a thin exine (about 0.8 μm) but a thick electron-transparent intine (2.5 μm) crossed by channels containing proteins of gametophytic origin.

Cytoplasm (Figure 3.7) is mainly rich in lipid bodies and vesicles. Plastids are less evident, whereas numerous mitochondria and fragmented endoplasmic reticulum appear. Mature pollen is bicellular with a thin, elongated generative cell and a roundish small vegetative nucleus (Figure 3.8).

Among the normally structured pollen grains the anomalous ones are numerous. These are generally smaller, empty, collapsed, and their wall broken in many places. An anther can contain up to 74% anomalous grains. Pollen viability (Table 3.1), assayed by means of fluorescein diacetate (FDA^+), is about 66%, whereas alcohol dehydrogenase activity (ADH^+) is lower (57%). Aborted pollen grains revealed by the Alexander method (Alexander 1969) are always high in number (Figure 3.9, Table 3.1). This low viability has to be related to saffron's triploid genome (Mathew 1982). This anomalous composition in pollen grains results in low pollen germination both *in vitro* and *in vivo* (Table 3.2). Differences are not evident when pollen germination occurs on stigmas after self- or outcross pollination. Characteristics are the anomalies accompanying pollen germination, concerning the aspects and behaviour of the pollen-tube emission and growth. Bifurcate, enlarged, or thin tips, or spirally elongated pollen tubes, occur frequently in pollen germinated in media or on stigmas and in styles.

C. thomasii (Figures 3.10, 3.11) and *C. cartwrightianus* have pollen grains homogeneously shaped and sized but smaller than those of *C. sativus* (Table 3.1). Pollen

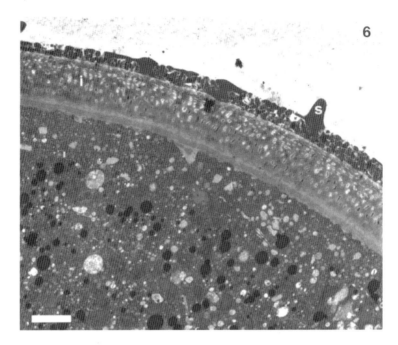

Figure 3.6 *Crocus sativus* pollen, TEM micrographs. Median section of pollen showing the external exine with spinulae (s) and outer intine (i) with channels containing proteins and multilayered inner intine. In the cytoplasm, lipid bodies and numerous vesicles are visible [Bar = 2.5 μm].

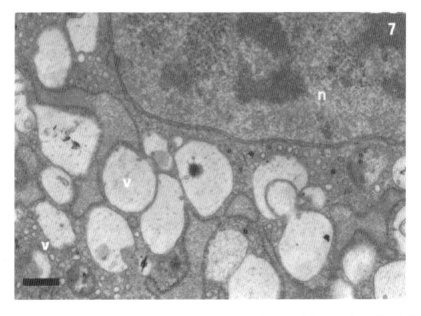

Figure 3.7 *Crocus sativus* pollen, TEM micrographs, Cytoplasm with large and small vesicles (v), endoplasmic reticulum (r) and vegetative nucleus (n) [Bar = 0.6 μm].

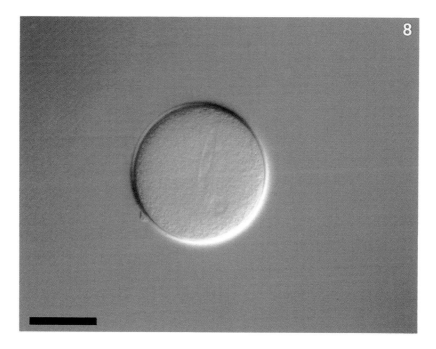

Figure 3.8 *Crocus sativus* pollen, Pollen grains of *C. sativus* under an interference microscope. Inside the cytoplasm, a roundish vegetative nucleus and spindle generative cell are visible [Bar = 50 μm].

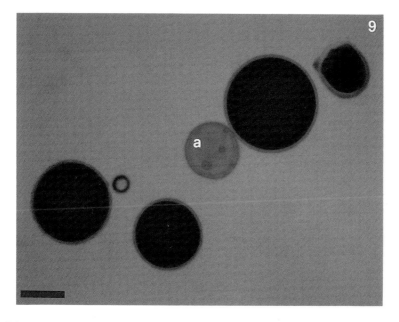

Figure 3.9 *Crocus sativus* pollen, Aborted (a) and viable pollen grains after Alexander staining under light microscopy [Bar = 50 μm].

Table 3.1 Pollen size (μm) and percentage of viable (FDA⁺, ADH⁺), anomalous and aborted pollen grains in *Crocus sativus*, *C. cartwrightianus* and *C. thomasii*

Crocus Species	Size (μm)	FDA⁺ (%)	ADH⁺ (%)	Abnormal Grains(%)	Aborted Grains (%)
C. sativus	92.1±15	66.0	56.8	34	31
C. cartwrightianus	63.8±8	87.3	79.2	12.7	15
C. thomasii	66.5±7	88.5	66	11.4	6

Table 3.2 Germinated pollen grains (%) of *Crocus sativus*, *C. cartwrightianus* and *C. thomasii* *in vitro* and after self- and cross-pollination

Pollen	In vitro	On Stigmas	
		Self-pollination	Cross-pollination
C. sativus	20	21	22
C. cartwrightianus	31	52	46
C. thomasii	–	58	56

ultrastructure shows the pollen wall, with the exine thinner than in *C. sativus*, whereas cytoplasmic organization is the same (Grilli Caiola and Di Somma 1994). All three species have bicellular pollen. The vegetative nucleus in *C. thomasii* is deeply lobed and surrounds the generative cell. The percentage of anomalous and aborted grains is low so a large amount of the pollen germinates, mainly after outcross-pollination. Crossed pollination between these three species shows that *C. sativus* pollen is highly sterile on the stigmas of *C. thomasii* and *C. cartwrightianus*, but the pollen of these last two species on stigmas of *C. sativus* reveals high compatibility with *C. thomasii*, but less with *C. cartwrightianus*, From this point of view, *C. sativus* and *C. cartwrightianus* seem

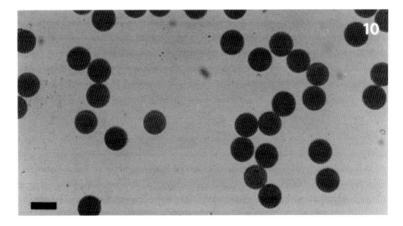

Figure 3.10 *Crocus thomasii* pollen grain after ADH treatment [Bar = 100 μm].

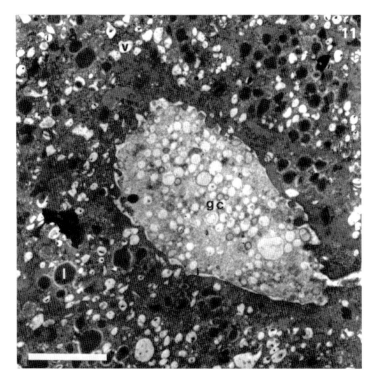

Figure 3.11 Details of vegetative cytoplasm and generative cell (gc) of *Crocus thomasii* pollen under EM. Lipid bodies (l) and vesicles (v) are visible in the vegetative cytoplasm [Bar = 1 μm].

to share common incompatibility genes. *C. thomasii* has the highest percentage of pollen germination on stigmas both of *C. sativus* and *C. cartwrightianus*, but much more so on the latter.

PISTIL ORGANIZATION

C. sativus pistil organization has been studied in flower buds and in flowers at different developmental stages (Grilli Caiola and Chichiriccò 1991). At anthesis, the saffron pistil has a dry-type stigma with papillae that are covered by a thick continuous cuticle. Stigmas are longer (about 2 cm) than the anthers and frequently also longer than the tepals. They are erect during anthesis but as the flower opens they bend downwards. The style is about 9 cm long, internally made up of three separate channels forming a single cavity in the main tract lined with a layer of secretory cells. This secretory layer extends down to the ovary where the stylar cavity opens into three locules.

The ovary is tricarpellar and trilocular. Along the axial region of the locules, placentas differentiate the ovules (Figure 3.12). The ovules are obliquely attached to the ovarian axis in six longitudinal rows, two for each locule, forming a total of 18–

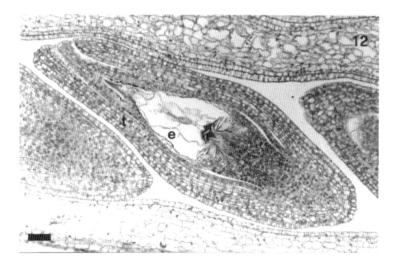

Figure 3.12 Longitudinal section of an ovule of *Crocus sativus* in which the integuments (t) and the embryo sac (e) with female gametophyte are visible [Bar = 10 μm].

20 ovules per locule. Ovarian septa widen towards the base and much more towards the style, narrowing at the top of the locule where there is a wide sterile portion.

Three symmetrically arranged canals traverse the ovarian septa longitudinally, originating approximately at the level of the median part of the ovary and ending at the base of the style. Located between two vascular bundles, they expand in the upper part of the ovary and cover about half the length of the septa. Ovarian canals are lined with an epidermal layer of radially elongated secretory cells, similar to septal nectaries present in other *Crocus* species. However, only a little secretion occurs. Ovules are anatropous and bitegmic with a large hypostasis (Figure 3.12). The external integument extends beyond the internal one and forms a narrow micropylar canal. Megasporogenesis occurs early upon sprouting in September and an embryo sac appears in the ovules of flower buds at the 1.5 to 2 cm long stage, when the flower is fully enveloped by cataphylls. There are therefore no differences between embryo sacs from floral buds and those from young and mature flowers. In fact, the embryo sac preserves its structure for some time after the wilting of the flowers. About 90% of the ovules develop an embryo sac which is seven-nucleate when mature. Most of the embryo sacs contain a substance that stains red with Poinceau 2R, specially during the initial developmental stages. None of the control saffron plants, unpollinated, freely or hand pollinated, developed fruits or seeds, indicating the absence of apomittic processes.

The pistils of *C. cartwrightianus* and *C. thomasii* are organized similarly to *C. sativus*, but the dimensions of the ovary, and the number of ovule and mature embryo sacs within it at anthesis, differ. The ovary of *C. cartwrightianus* contains 18–20 ovules per locule. The inner ovule integument is formed by four to five layers, that of the outer one by five to six. *C. thomasii* has 24–28 ovules per locule and each ovule has an inner integument of four to five layers at the micropyle and five to six at the outer wall.

In all three species, the megagametophyte differentiates early, being just evident in the flower bud. Then the embryo sac maintains its integrity until some days after anthesis.

FERTILIZATION

Studies on saffron pollen biology *in vitro* (Chichiriccò and Grilli Caiola 1982, 1984, 1986, Grilli Caiola *et al.* 1985) have indicated that a high percentage of the pollen is infertile (Table 3.2), but this need not be a barrier to seed set because each anther produces a high number of pollen grain, (about 7,000). Problems arise when pollen germinates on the stigma (Figure 3.13), after both self- and outcross pollination. Pollen germination on the stigma indicated that saffron is self- and outcross sterile, but pollen from *C. thomasii* or *C. cartwrightianus* is able to germinate (Table 3.3) and grow in the saffron pistil. Fertilized ovules and fruit set have been obtained *in vitro* from *C. sativus* pistils pollinated with *C. thomasii* pollen (Chichiriccò 1989c). Parthenocarpic fruit development from ovaries cultured on an agar medium supplemented with growth substances was independent of ovary age (before, during and after

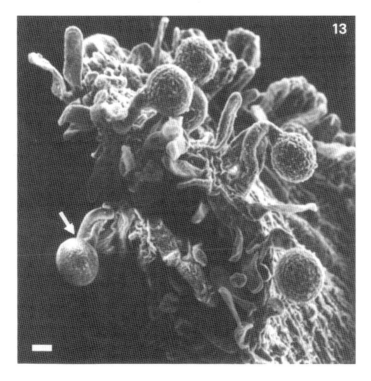

Figure 3.13 Germinated pollen grains (arrow) on *Crocus sativus* stigma 2 h after cross-pollination [Bar = 50 µm].

Table 3.3 Percentage of germinated pollen grains on stigmas after pollination between *Crocus sativus, C. cartwrightianus* and *C. thomasii*

On Stigma of	Germinated Pollen (%) of		
	C. sativus	*C. cartwrightianus*	*C. thomasii*
C. sativus	–	28	59.2
C. cartwrightianus	20	–	72
C. thomasii	22	50	–

anthesis), confirming that ovules remain viable a long time after anthesis (Chichiriccò and Grilli Caiola 1987, Grilli Caiola and Chichiriccò 1991). Self- and cross-pollination of saffron also revealed that it does not produce seeds after intraspecific pollination, but capsules and seeds mature after stigmatic pollination with *C. thomasii* pollen (Chichiriccò 1989c). Both *C. thomasii* and *C. cartwrightianus* are self-sterile but outcross-fertile species (Grilli Caiola 1994, 1995).

Experiments carried out in the field led to the finding that although self-, outcross- and unpollinated pistils of saffron did not produce fruits and seeds, a capsule was obtained from a saffron plant grown near *C. cartwrightianus* plants (unpublished data).

In nature, this is possible where *C. sativus* and *C. cartwrightianus* flower simultaneously, and the weather is hot (about 25°C) and sunny. These ecological conditions favour the presence of visiting pollinators which are attracted by the scented *Crocus* flowers. Observations carried out for some years led to the identification of the hymenopteran *Bombus silvestris* as responsible for interspecific pollinations between saffron and *C. cartwrightianus*. In fact, *Bombus* appears in groups of 2 to 10 individuals during the late morning hours, when flowers open and emanate their scent. Under these conditions, insects collect pollen from numerous flowers. Because they do not discriminate between different species of pollen, such as *C. sativus* and *C. cartwrightianus*, pollination between different species is possible.

The capsules and seeds obtained were larger than those of *C. cartwrightianus* and *C. thomasii*, but very similar in other aspects such as shape, colour, seed arrangement and capsule dehiscence.

THE ROLE OF CALCIUM ION IN THE REPRODUCTION PROCESS OF SAFFRON AND ITS ALLIES

The role of calcium in flower plant reproduction has been reported in a number of processes. Ca^{2+} ions affect processes such as pollen germination, pollen-tube growth, and control of pollen germination on stigma in incompatible sporophytic processes, (Bednarska 1989, Frankling-Tong *et al.* 1993). Also, in the fertilization processes, Ca^{2+} has been thought to be involved in the chemotropic guidance of the pollen tube towards the region of synergids and to be the cause of the tip's rupture for the release of sperm cells (Chaubal and Reger 1994).

Saffron pollen germination *in vitro* shows some anomalies of pollen-tube emission and growth similar to those reported in other plants when pollen germinates in a calcium-free medium (Brewbaker and Kwack 1963, Pfahler 1967).

In saffron and its allies calcium ions do not seem to be necessary to start germination *in vitro*, but appear to be related to regular growth of the pollen tube both *in vitro* and *in vivo*. Calcium-ion concentration in the stigmas, styles and ovaries of unpollinated and self- and cross-pollinated pistils of *C. sativus* and *C. cartwrightianus* has been detected by means of calcium-selective microelectrodes (Brandizzi and Grilli Caiola 1996b). Results indicated that calcium-ion concentration decreases in unpollinated and pollinated infertile pistils of *C. sativus*, whereas in the fertile ovaries of *C. cartwrightianus*, an increase in this element appears in relation to ovule fertilization after stigmatic cross-pollination. Thus increased calcium levels in the fertilized ovaries can be a signal that fertilization has occurred. Similar results have been reported in the self- and cross-pollinated fertile *Crocus biflorus* (in press) and in the cross-pollinated fertile ovary of the iridacean *Hermodactylus tuberosus* L. (Grilli Caiola and Brandizzi 1994).

Ultrathin pollen sections observed by EM by electron spectroscopy imaging (ESI) and electron energy-loss spectroscopy (EELS) (Grilli Caiola *et al.* 1996) revealed a high calcium content in the pollen wall as well as in the cytoplasm of *C. sativus* and *C. cartwrightianus*, confirming that calcium is not a limiting factor for pollen germination in saffron. Fluorescence detected by chlorotetracycline treatment in pollen grains and papillae before and after pollination revealed a constant calcium decrease in pollen grain and in papillae in both unpollinated and pollinated stigmas. All these results suggest that calcium is involved in the fertilization processes of fertile *Crocus* species. Detection of calcium concentration in the pollen and pistil of *C. sativus*, *C. cartwrightianus* and *C. thomasii* at different developmental stages (bud flower, before and at anthesis) indicated that calcium accumulates in the upper parts of the pistil from the flower bud stage to anthesis. The *C. thomasii* pistil is richest in Ca^{2+}. It reaches the highest values of 2.3×10^{-3} M $\pm$ 0.1 in the stigma of the closed flower, whereas the lowest (5.5×10^{-6} M) and constant values have been obtained in the pistil of closed *C. sativus* flowers.

Although these data do not enable one to draw conclusions about the mechanisms regulating the compatibility and incompatibility processes, and infertility and fertility in *C. sativus* and its allies, they do indicate that calcium ions can be one of numerous factors regulating seed and fruit set.

REFERENCES

Alexander, M.P. (1969) Differential staining of aborted and nonaborted pollen. *Stain Technology*, **44**, 117–122.

Bednarska, E. (1989) The effect of exogenous Ca^{2+} ions on pollen grain germination and pollen tube growth. Investigations with $^{45}Ca^{2+}$ together with Verapamil, La^{3+}, and ruthenium red. *Sex. Plant Reprod.* **2**, 53–58.

Brandizzi, F. and Grilli Caiola, M. (1996a) Quantitative DNA analysis in different *Crocus* species (Iridaceae) by means of flow cytometry. *Giornale Botanico Italiano* **130**, 643–645.

Brandizzi, F. and Grilli Caiola, M. (1996b) Calcium variation in pistil of *Crocus cartwrightianus* Herb. and *Crocus sativus* L. *J. Trace and Microprobe Techniques*, **14**, 4115–4126.

Brewbaker, J.L. and Kwack, B.H. (1963) The essential role of calcium ion in pollen germination and pollen tube growth. *American Journal of Botany*, **50**, 859–865.

Brighton, C.A. (1977) Cytology of *Crocus sativus* and its Allies (Iridaceae). *Plant Systematics and Evolution*, **128**, 137–157.

Chaubal, B.J. and Reger, J.B. (1994) Dynamics of antimonate-precipitated calcium and degeneration in unpollinated pearl millet synergids after maturity. *Sex. Plant Reprod.*, **7**, 122–134.

Chichiriccò, G. (1987) Megasporogenesis and development of embryo sac in *Crocus sativus* L. *Caryologia*, **40**, 59–69.

Chichiriccò, G. (1989a) Embryology of *Crocus thomasii* (Iridaceae). *Plant Systematics and Evolution*, **168**, 39–47.

Chichiriccò, G. (1989b) Microsporogenesis and pollen development in *Crocus sativus* L. *Caryologia*, **42**, 237–249.

Chichiriccò, G. (1989c) Fertilization of *Crocus sativus* L. ovules and development of seed after stigmatic pollination with *C. thomasii* Ten. pollen. *Giornale Botanico Italiano*, **123**, 31–37.

Chichiriccò, G. (1996) Intra- and interspecific reproductive barriers in *Crocus* (Iridaceae). *Plant Systematics and Evolution*, **201**, 83–92.

Chichiriccò, G. and Grilli Caiola, M. (1982) Germination and viability of the pollen of *Crocus sativus* L. *Giornale Botanico Italiano*, **116**, 167–173.

Chichiriccò, G. and Grilli Caiola, M. (1984) *Crocus sativus* pollen tube growth in intra- and interspecific pollinations. *Caryologia*, **37**, 115–125.

Chichiriccò, G. and Grilli Caiola, M. (1986) *Crocus sativus* pollen germination and pollen tube growth *in vitro* and after intraspecific and interspecific pollination. *Canadian Journal of Botany*, **64**, 2774–2777.

Chichiriccò, G. and Grilli Caiola, M. (1987) *In vitro* development of parthenocarpic fruits of *Crocus sativus* L. *Plant Cell Tissue and Organ Culture*, **11**, 75–78.

Franklin-Tong, V.E., Ride, J.P., Read, N.D., Trewavas, A.J. and Franklin, F.C.H. (1993). The self-incompatibility response in *Papaver rhoeas* is mediated by cytosolic free calcium. *Plant J.*, **4**, 163–177.

Grilli Caiola, M. (1994) Pollen structure and germination of *Crocus thomasii* Ten. (Iridaceae). *Giornale Botanico Italiano*, **128**, 869–877.

Grilli Caiola, M. (1995) A study on pollen grains of *Crocus cartwrightianus* (Iridaceae). *Plant Systematics and Evolution*, **198**, 155–166.

Grilli Caiola, M. and Brandizzi, F. (1994) Pistil calcium content and pollen germination in *Hermodactylus tuberosus* (L.) Mill. (Iridaceae). *Giornale Botanico Italiano*, **128**, 70.

Grilli Caiola, M., Brandizzi, F. and Canini, A. (1996) Calcium localization in pollen of *Hermodactylus tuberosus* Mill. (Iridaceae). *Giornale Botanico Italiano*, **130**, 400.

Grilli Caiola, M. and Chichiriccò, G. (1991) Structural organization of the pistil in saffron (*Crocus sativus* L.). *Israel Journal of Botany*, **40**, 199–207.

Grilli Caiola, M., Castagnola, M. and Chichiriccò, G. (1985) Ultrastructural study of Saffron (*Crocus sativus* L.) pollen. *Giornale Botanico Italiano*, **119**, 61–66.

Grilli Caiola, M. and Di Somma, D. (1994) Comparative study on pollen of different *Crocus sativus* aggregatus species. *13th Int.* Congress *ISSPR*, Wien, July 1994.

Igarashi, Y., Hisada, A. and Yuasa, M. (1993) Regeneration of flower buds in culture of ovaries of saffron (*Crocus sativus* L.). *XV International Botanical Congress*, Tokyo.

Karasawa, K. (1933) On the triploidy of *Crocus sativus* L. and its high sterility. *Japanese Journal of Genetics*, **9**, 6–8.

Mathew, B. (1982) *The Crocus: a Revision of the Genus Crocus (Iridaceae)*. Batsford, London.

Pflahler, P.L. (1967) *In vitro* germination and pollen tube growth of maize (*Zea mays* L.) pollen. I. Calcium and boron effects. *Canadian Journal of Botany*, **45**, 839–945.

4. SAFFRON CHEMISTRY

DOV BASKER

Department of Food Science, Agricultural Research Organization,
The Volcani Centre, Bet Dagan, Israel

ABSTRACT   The principal components of saffron are discussed, those responsible for its colour, odour and taste. Analysis and any possible toxicity are also mentioned.

The proximate analysis of commercial saffron – the dried red stigmas of *Crocus sativus* L. – has been reported (Nicholls 1945, Sastry *et al.* 1955, Triebold and Aurand 1963, Stecher 1968, Indian Standard 1969, International Standards Organization 1970, Melchior and Kastner 1974, Sampathu *et al.* 1984, Basker and Negbi 1985, Skrubis 1990) to give data, in % w/w, as in Table 4.1. The problems that precede chemical analysis include the guarantee of correct botanical identification, the risk of partial adulteration, and the presence of floral waste. It is probably inevitable that parts of the yellow-to-uncoloured style, as well as anthers and possibly some petals or even leaves, are found.

Various limits are set for the quantity of floral waste (Nicholls 1945, International Standards Organization 1970, Krogh and Akenstrand 1980) to below given levels (1%, 5%, 10%, 15%), depending on the declared quality category. While leaves should really not be present at all, flowers picked once they have begun to wilt after their 2- to 3-day bloom cannot readily be separated into their constituent parts (Basker 1993) (see the chapter on saffron technology), and separate analysis of styles, particularly their tops (Skrubis 1990), may yet indicate the presence of uncoloured positive-quality taste parameters.

The most obvious characteristic of saffron is its deep red colour. The high gloss of fresh stigmas becomes dulled upon drying, and a strong yellow extract passes into water upon wetting the dried stigmas. Colour intensity is expressed as the extinction (= optical density) of a hypothetical 1% (w/v) solution at the wavelength of its visible spectral maximum, in a 1-cm cell, and is written as $E_{1cm}^{1\%}$ (Booth, 1957:9). The International Standards Organization (1980a) specified that $E_{1cm}^{1\%}$ for Category I commercial saffron be not less than 110 or 150 for hay and powdered saffron – whose colour is presumably extracted more completely –, respectively (see the chapter on saffron technology), in water at 440 nm. For comparison, pure β-carotene at 451 nm in cyclohexane has an $E_{1cm}^{1\%} = 2505$ (Issler and Schudel 1963). The precise wavelength at the spectral peak of carotenoids tends to shift as a function of the solvent employed (Booth 1957:6), and this must therefore always be stated.

The colour of saffron is due principally to a water-soluble carotenoid, α-crocin. The structures of some simple, water-insoluble carotenoids are shown (in Figure 4.1) for simplicity in the all-*trans* forms; their visible colour is due to the conjugate (alternating single with) double bonds. Thus phytofluene (Figure 4.1) with five conjugated

Table 4.1 Proximate analysis of commercial saffron (% w/w)

Moisture	10
Water-soluble matter	53
including Sugars (as invert)	14
Gums	10
Pentosans	8
Pectin	6
Starch	6
α-Crocin	2
Other carotenoids	1
Protein (N * 6.25)	12
Inorganic matter (ash)	6
including Ash insoluble in HCl	0.5
Nonvolatile oils	6
Volatile oils	1
Crude fibre	5

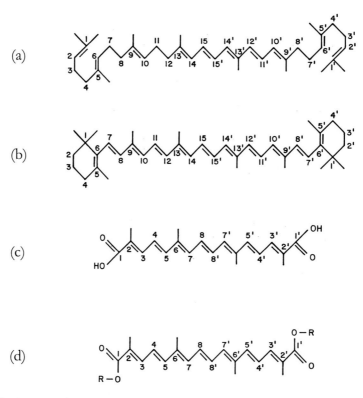

Figure 4.1 Structural formulae of some carotenoids. (a) phytofluene, $C_{40}H_{68}$, 5 conjugated double bonds. (b) β-carotene, $C_{40}H_{56}$, 11 conjugated double bonds. (c) crocetin, $C_{20}H_{24}O_4$, 9 conjugated double bonds. (d) α-crocin, $C_{44}H_{64}O_{24}$, 9 conjugated double bonds. R: gentiobiose (see Figure 4.2), linked in α-configuration.

double bonds is practically colourless (Zechmeister, 1962) while phytoene, with only three conjugated double bonds – it has a single bond in the 11–12 position – has no colour at all (*ibid.*). On the other hand, the strongly coloured and most abundant natural carotenoid (Booth, 1957:7), β-carotene (Figure 4.1) has 11 conjugated double bonds; the structure of lycopene (found in tomatoes, *etc.*) is similar to that of β-carotene except that the 1–6 and 1'–6' positions are not bound to one another, and double bonds are present instead at the 1–2 and 1'–2' positions (*cf.* phytofluene). The structure of α-crocin is based on the shorter carbon chain of crocetin (Figure 4.1) which has nine conjugated double bonds; α-crocin (Figure 4.1) is also a glycoside, a type of compound with a sugar – in this case gentiobiose (Figure 4.2) – and it is the presence of these moieties at both ends of the carotenoid molecule which account for its water-solubility. Other carotenoids, in both water-soluble and -insoluble forms, are present in lower concentrations (Pfander and Wittwer 1975 a, b, Dhingra *et al.* 1975).

A principal aim of commercial saffron analysis is the determination of the extractable colour intensity on either macro- or micro-samples, gram or milligram quantities, respectively (International Standards Organization 1980b, Hanson 1973, Basker and Negbi 1985). Because of the risk of adulteration (Khanna *et al.* 1980, Kapur 1988), this is generally followed by thin-layer chromatography (TLC) of the aqueous extract to separate colour components, preferably with saffron of assured purity as a control (International Standards Organization 1970, Foppen 1971, Parvaneh 1972, Zweig and Sherma 1972, Dhar and Suri 1974, French Standard 1976).

Because of the high value of saffron (Basker 1993), many other types of adulteration have been attempted by unscrupulous dealers (*e.g.*, Pliny, 1st century CE, Lowell 1964, *Encyclopaedia Judaica* 1973, Sampathu *et al.* 1984), even at the risk of capital punishment (Bowles 1952, Meyer 1982).

All carotenoids are subject to oxidation, C = C bonds opening to receive oxygen atoms and consequently diminishing the characteristic colour. The reaction is catalysed by light. The conjugative nature of the bond chain provides some protection from oxidation, but on the other hand, the reaction protects an organism from further oxidative damage.

The undried fresh stigmas have a strong attractive odour to humans as well as to bees. Bees found in crocus flowers are frequently soporific. The odour of commercially dried saffron has been identified as principally due to an aldehyde, safranal (Figure 4.2), but it is not known whether the identity holds for the fresh odour as well. Safranal boils at 172°C at atmospheric pressure (Furia and Bellanca 1975), and is sufficiently volatile at lower temperatures to be lost if given enough time (*ibid.*, Guenther 1952). Other odoriferous volatiles are present as well, in concentrations varying from 2 to 29% relative to safranal (Zarghami and Heinz 1971).

The bitterish but pleasant taste of saffron (Sastry *et al.* 1955) is due to a glucoside (a glycoside with glucose as the sugar), picrocrocin (Figure 4.2), which can be broken down by heating (Stahl and Wagner 1969) or enzymatically (Guenther 1952, Zarghami 1970) into safranal and D-glucose (Figure 4.2). "Very fresh" saffron is reported to contain *ca* 4% picrocrocin (Guenther 1952), also known as saffron-bitter (Parry 1962, Stecher 1968). Figure 4.2 also shows the structural difference between the β- and α-configurations of D-glucose (and other sugars): the β of β-carotene has an entirely different connotation.

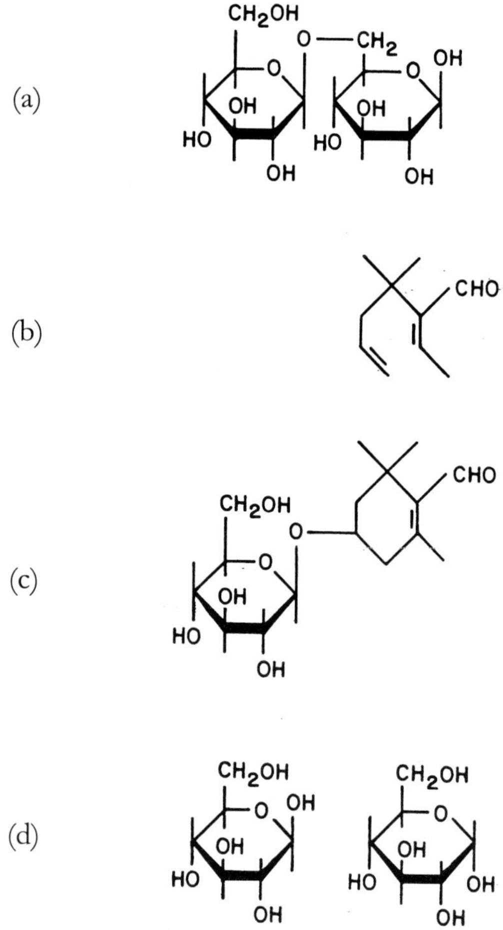

Figure 4.2 Structural formulae of some other compounds in saffron. (a): gentiobiose, $C_{12}H_{22}O_{11}$, (see Figure 4.1): β, 1–6' link between glucose rings. (b) safranal, $C_{10}H_{14}O$. (c) picrocrocin, $C_{16}H_{26}O_7$. (d): D-glucose, $C_6H_{12}O_7$, 6-membered pyranose ring form. Left: β-D-glucose. Right: α-D-glucose.

Once it has been established that a sample of saffron is authentic, unadulterated, and free of excessive waste material, its commercial value relative to the market at the moment (Basker 1993) depends on further quality characteristics. For many consumers who use it only as a magnificent yellow food colour (Sastry et al. 1955, Rietz 1961, Zarghami 1970), this parameter is quantitatively and readily determined spectrophotometrically as above. Quantitative and simple determinations of safranal and picrocrocin, even approximate, are more problematic. Saffron's aqueous spectrum shows three peaks of different heights, at about 440 (visible), 325 and 255 nm (ultraviolet) (Basker and Negbi 1985). It has been reported that the two ultraviolet peaks can be used to estimate relative concentrations of safranal and picrocrocin, respectively, either directly (International Standards Organization 1990) or by difference from a low-point at 297 nm (Corradi and Micheli 1979). Confirmation of such methods would be desirable and useful: the pure products are required for calibration purposes – methods for their laboratory extraction from saffron are given by Guenther (1952) (safranal by steam-distillation), by Kuhn and Winterstein (1934) (picrocrocin by solvent extraction), and by Iborra *et al.* (1992) for microsamples. HPLC (high-performance liquid chromatography) has also been employed (Solinas and Cichelli 1988).

The variable composition of saffron is a drawback for pharmaceutical purposes, compounded by the absence of quantitative experimental evidence on the effects of its major individual components. Before the scientific era, saffron was used, among other purposes (Basker and Negbi 1983), for the treatment of cardiac, lung, digestive and feminine disorders (Dioscorides, 1st century CE, Pliny, 1st century CE, Maimonides, 12th century CE, Gerard 1633, Buley 1933, Dawson 1934, Warren 1970, Lewis and Elvin-Lewis, 1977). Although saffron has now been reduced therapeutically to the status of a herbal remedy (Folch Andreu 1957, Bailey 1975, Lust 1978), it is difficult to dismiss medical experience out of hand (Sexton 1950, Gainer and Chisolm 1974, Grisolia 1974, Nishio *et al.* 1987, Panikkar 1990). The dose is also a matter of dispute, probably arising from the wide composition-range of the commercial product. This is of some importance, as anything in a large enough dose can be toxic (Stevens and Klarner 1990). Maimonides (12th century CE) warned that saffron "in excess" depresses appetite, the *Encyclopaedia Britannica* (1974) that unspecified "overdoses" are narcotic. Gerard (1633) and Culpeper (1652) advised about 0.5 g "for those at death's door" and 1.5 g has proved fatal (Fasal and Wachner 1933), although Lust (1978) warns only about a 10 g dose; yet a trial-by-ordeal with such a dose in 16th century CE India resulted in survival, and thus freedom, for the accused (Holkar and Holkar 1975). Arena (1974) on one page describes the effects of poisoning by saffron, and on another states that it has no toxic effects – possibly the results of quite different doses.

REFERENCES

Arena, J.M. (1974) *Poisoning: Toxicology, Symptoms, Treatments* (3rd edn). Charles C. Thomas, Springfield, IL, pp. 463, 592.
Bailey, L.H. (1975) *Manual of Cultivated Plants*. Macmillan Publishing Co., Inc., New York, p. 265.
Basker, D. (1993) Saffron, the costliest spice: drying and quality, supply and price. *Acta Hort.*, **344**, 86–97.
Basker, D. and Negbi, M. (1983) Uses of saffron. *Econ. Bot.*, **37**, 227–235.
Basker, D. and Negbi, M. (1985) Crocetin equivalent of saffron extracts: comparison of three extraction methods. *J. Assoc. Publ. Analysts*, **23**, 65–69.
Blacow, N.W. (Ed.) (1972) *Martindale, The Extra Pharmacopoeia* (26th ed.). The Pharmaceutical Press, London, p. 726.
Booth, V.H. (1957) *Carotene; Its Determination in Biological Materials*. Heffer, Cambridge.
Bowles, E.H. (1952) *A Handbook of Crocus and Colchicum*, The Bodley Head, London, pp. 63.
Buley, R.C. (1933) Pioneer health and medical practices in the old northwest prior to 1840. *Mississippi Valley Historical Review*, **20**, 497–520.

Corradi, C. and Micheli, G. (1979) Determinazione spettrofotometrica del potere colorante, amaricante ed odoroso dello zaferano. *Boll. Chim. Farm*, **118**, 553–562.

Culpeper, N. (1652) Quoted by Silberrad and Lyall (1909).

Dawson, W.R. (1934) *A Leechbook or Collection of Medical Recipes of the Fifteenth Century*, Macmillan, London, p. 71.

Dhar, D.N. and Suri, S.C. (1974) Thin layer chromatographic detection of dyes as adulterants in saffron. *J. Inst. Chemists (India)*, **46**, 130–132.

Dhingra, V.K., Seshadri, T.R. and Mukerjee, S.K. (1975) Minor carotenoid glycosides from saffron (*Crocus sativus*). *Indian J. Chem*, **3**, 339–341.

Dioscorides (1st century CE) *The Greek Herbal*, Goodyer, J. (translator, 1655), Gunther, R.T. (Ed., 1933), reprinted 1959. Hafner, New York, p. 22.

Encyclopaedia Britannica (1974) Encyclopaedia Britannica Inc., Chicago, IL, Macropaedia, Vol. **9**, p. 891.

Encyclopaedia Judaica (1973) Keter Publishing House, Jerusalem, Vol. **14**, pp. 631.

Fasal, P. and Wachner, G. (1933) *Wein. klin. Wschr.*, **45**, 745 Quoted by Blacow (1972).

Folch Andreu, R. (1957) [A drug which is disappearing from the medical thesaurus: saffron. An historical study.] *Farmacognosia*, **17**, 145–224 (in Spanish).

Foppen, F.H. (1971) Tables for the identification of carotenoid pigments. *Chromatogr. Rev.*, **14**, 133–298.

French Standard (1976) [*Spices and Condiments. Saffron: Identification of Pigments.*], NFV 32–124 (in French).

Furia, T.E. and Bellanca, N. (Eds.) (1975) *Fenaroli's Handbook of Flavor Ingredients*. CRC Press, Inc., Cleveland, OH, Vol **2**, p. 515.

Gainer, J.W. and Chisolm, G.M. (1974) Oxygen diffusion and atherosclerosis. *Atherosclerosis*, **19**, 135–138.

Gerard, J. (1633) *The Herbal or General History of Plants*, reprinted 1975. Dover Publications Inc., New York, p. 154.

Grisolia, S. (1974) Hypoxia, saffron, and cardiovascular disease. *Lancet*, **2**, 41–42.

Guenther, E. (1952) *The Essential Oils*. Van Nostrand Company, Inc., New York, Vol. **6**, p. 105.

Hanson, N.W. (Ed.) (1973) *Official, Standardised and Recommended Methods of Analysis* (2nd edn). Society for Analytical Chemistry, London, p. 674.

Holkar, S.R. and Holkar, S.D. (1975) *Cooking of the Maharajas*. Viking Press, Inc., New York, p. 249.

Iborra, J.L., Castellar, M.R., Canovas, M. and Manjon, A. (1992) TLC preparative purification of picrocrocin, HTCC and crocin from saffron. *J. Food Sci.*, **57**, 714–716, 731.

Indian Standard (1969) *Specification for Saffron*. Indian Standards Institution, IS: 5453–1969.

International Standards Organization (1970) *Saffron*, 3rd draft proposal. ISO/TC 34/SC 7, 215E, Geneva.

International Standards Organization (1980a) *Saffron: Specification*. ISO 3632, Geneva.

International Standards Organization (1980b) *Spices and Condiments – Determination of Cold Water Extract*. ISO 941, Geneva.

International Standards Organization (1990) *Spices and Condiments – Saffron – Test Methods*. Draft International Standard ISO/DIS 3632-2, Geneva.

Issler, O. and Schudel, P. (1963) Synthese und Markierung von Carotinen und Carotinoiden. In K. Lang (Chairman), *Carotine und Carotinoide*, Symposium, Wissenschaftliche Veroffentlichungen der Deutschen Gesellschaft fur Ernahrung, Mainz, October 1961, Steinkopff Verlag, Darmstadt.

Kapur, B.M. (Ed.) (1988) *Annual Report*, Regional Research Laboratory, Council of Scientific and Industrial Research, Jammu, pp. 59–60.

Khanna, S.K., Singh, G.B. and Krishnamurti, C.R. (1980) Toxicity profile of some commonly encountered food colours. *J. Food Sci. Tech. (India)*, **17**, 95–103.

Krogh, G. and Akenstrand, K. (1980) [Saffron – authentic or adulterated?]. *Var Foda*, **33**, 346–353 (in Swedish).

Kuhn, R. and Winterstein, A. (1934) Uber die Konstitution des Pikro-crocins und seine Beziehung zu den Carotin-Farbstoffen des Safrans. *Ber. dtsch. chem. Ges.*, **67**, 344–357.
Lewis, W.H. and Elvin-Lewis, M.P.F. (1977) *Medical Botany.* J. Wiley & Sons, New York, pp. 325, 329.
Lowell, G. (1964) Saffron adulteration. *J. Assoc. Offic. Agric. Chemists*, **47**, 538.
Lust, J.B. (1978) *The Herb Book*. Bantam Books Inc., New York, p. 341.
Maimonides, M. (12th century CE) *On the Causes of Symptoms*, Leibowitz, J.O. and Marcus, S. (Eds.) (1974). Univ. Calif. Press, Berkeley, CA, p. 125.
Melchior, H. and Kastner, H. (1974) *Gewurze*. Verlag Paul Parey, Berlin, pp. 147–150.
Meyer, G.L. (1982) Knossus: world trade centre in saffron. *Organorama*, **19**, 11–14.
Nicholls, J.R. (1945) *Aid to the Analysis of Food and Drugs* (6th edn). Bailliere, Tindall and Cox, London, p. 195.
Nishio, T., Okugawa, H., Kato, A., Hashimoto, Y., Matsumoto, K. and Fujioka, A. (1987) [Effect of crocus (*Crocus sativus* Linne, Iridiaceae) on blood coagulation and fibrinolysis.] *Shoyakugaku Zasshi*, **41**, 271–276 (in Japanese, with English abstract).
Panikkar, K.R. (1990) Antitumour activity of saffron. *Cancer Letters*, **57**, 109–114.
Parry, J.W. (1962) *Spices: Their Morphology, Histology and Chemistry*. Chemical Publishing Co., Inc., New York, p. 208.
Parvaneh, V. (1972) A note on the assessment of purity of saffron colour. *J. Assoc. Publ. Analysts*, **10**, 31–32.
Pfander, H. and Wittwer, F. (1975a) Untersuchungen zur Carotinoid-Zusammensetzung im Saffran II. *Helv. Chim. Acta*, **58**, 1608–1620.
Pfander, H. and Wittwer, F. (1975b) Untersuchungen zur Carotinoid-Zusammensetzung im Saffran III. *Helv. Chim. Acta*, **58**, 2233–2236.
Pliny (the Elder) (1st century CE) *Natural History*. Bostock, J. and Riley, H.T. (trans.) (1887), Bohn, London.
Rietz, C.A. (1961) *A Guide to the Selection, Combination and Cooking of Foods*. Avi Publishing Company, Inc., Westport, CT, Vol. **1**, p. 314.
Sampathu, S.R., Shivashankar, S. and Lewis, Y.S. (1984) Saffron (*Crocus sativus* Linn.) – cultivation, processing, chemistry and standardization. *CRC Crit. Rev. Food Sci. Nutr.*, **20**, 123–157.
Sastry, L.V. L., Srinivasan, M. and Subrahmanyan, V. (1955) Saffron (*Crocus sativus* Linn.) *J. Sci. Ind. Res. (India)*, 14–**A**, 178–184.
Sexton, W.A. (1950) *Chemical Constitution and Biological Activity*. Van Nostrand Company Inc., New York, p. 243.
Silberrad, U. and Lyall, S. (1909) *Dutch Bulbs and Gardens*. Adam and Charles Black, London, p. 38.
Skrubis, B. (1990) The cultivation in Greece of *Crocus sativus* L. In F. Tammaro and L. Marra (Eds.), *Lo Zafferano*. Proc. Internat. Conf. on Saffron, L'Aquila, Italy, Universita Della Studi L'Aquila e Accademia Italiana delli Cucina, L'Aquila, pp. 171–182.
Solinas, M. and Cichelli, A. (1988) [Verification of saffron quality and genuineness characteristics by HPLC analysis of colour and flavour components.] *Industrie Alimentari*, **27**, 634–639, 648 (in Italian, with English summary).
Stahl, E. and Wagner, C. (1969) TAS-method for the microanalysis of important constituents of saffron. *J. Chromatog.*, **40**, 308.
Stecher, P.G. (Ed.) (1968) *The Merck Index* (8th edn). Merck & Co., Inc., Rahway, NJ, pp. 928, 831.
Stevens, S.D. and Klarner, A. (1990) *Deadly Doses*. Writer's Digest Books, Cincinnati, OH.
Triebold, H.O. and Aurand, L.W. (1963) *Food Composition and Analysis*, Van Nostrand Company, Inc., Princeton, NJ, p. 463.
Warren, C.P.W. (1970) Some aspects of medicine in the Greek Bronze Age. *Med. Hist.*, **14**, 364–377.
Zarghami, N.S. (1970) *The Volatile Constituents of Saffron (Crocus sativus* L.). Ph.D Thesis, Univ. Calif. Davis, pp. 83.

Zarghami, N.S. and Heinz, D.E. (1971) Monoterpene aldehydes and isophorone-related compounds of saffron. *Phytochemistry*, **10**, 2755–2761.

Zechmeister, L. (1962) *Cis-trans Isomeric Carotenoids Vitamins A and Arylpolyenes*. Springer-Verlag, Vienna, pp. 102, 3, 114.

Zweig, G. and Sherma, J. (Eds.) (1972) *Handbook of Chromatography*. CRC Press, Cleveland, OH, Vol. **1**, pp. 540.

5. SAFFRON (*CROCUS SATIVUS* L.) IN ITALY

FERNANDO TAMMARO

*Department of Environmental Sciences, University of L'Aquila,
Via Vetoio, 67100 L'Aquila, Italy*

ABSTRACT   In Piano di Navelli (L'Aquila region, Central Italy), saffron is cultivated in annual cycles. There has been a great decrease in saffron production in this region over the past few years. Nevertheless, it represents a remarkable income for some farmers in this mountainous area with a poor economy. This arid area is atypical for saffron cultivation, but the unique annual cultivation results in the recognized superior quality of its saffron in the marketplace. This chapter describes the major aspects of saffron production in this region.

INTRODUCTION

According to tradition, a certain monk from Navelli (L'Aquila, Central Italy), on his return from Spain some time during the 15th century, adapted Spanish cultivation practices to the climate and soil of his village, in particular the development of cultivation in annual cycles. It is this practice, particular to Navelli and the Aquila area, that differs from those used in other countries (Spain, Greece, India, Sardinia, etc.) where the saffron plants are left in the soil from three to eight years (pluriannual cultivation) (Tammaro 1990). In this way, saffron cultivation which was well known among the ancient Roman people, but forgotten during the medieval age, was reintroduced in Italy.

Every year in Navelli the corms are taken up at the beginning of the summer and replanted at the end of August, after they have been selected for size and checked for possible defects (rot, parasites, viruses, etc.). The continual selection for size and checks for wholesomeness mean that every year only the best plants are replanted, and as a result only the highest morphological and phytochemical characteristics are conserved. This is why L'Aquila saffron is the most sought-after and most highly prized in the world.

It is interesting to note that during the last century in the L'Aquila area, experiments in pluriannual cultivation were carried out (the saffron was kept in the soil for three consecutive years). The pluriannual saffron plants were attacked earlier and more severely by root rot every year, promoting the resumption of annual cultivation.

At present, saffron in Italy is cultivated mostly in the highlands of Navelli, near L'Aquila (Central Italy); a few cultures can be found in Sardinia, Cagliari Province (Picci 1987) and in the Val di Taro, Parma Province (Zanzucchi 1987). In the past, saffron was widespread in many regions of Central and Southern Italy (Tuscany, Campania, Sicily, etc.), where it is no longer cultivated[1] for social (neglected counties), economic (low income) and biological (corm parasites) reasons.

Nowadays (1992–1995), the saffron cultivation area in Italy covers about 10 ha, much less than in the past (about 1000 ha). The total production (1994–1995) of dried saffron is about 70–80 kg, that of corms 18 tons. The price of dried stigmas reached US$4 per g. However, the saffron is packaged in small, artistic ceramic vases, and then sold at a price of about US$ 10 for 1.5 g. For the small number of saffron cultivators (about 100 people), these prices represent a remarkable income: the cultivators are located in sub-mountain areas with a poor economy.

Biological Cycle

The saffron plant is characterized by a biological cycle with a long pause in the summer and an active growth period in the autumn (also the period during which the flowers blossom). There is also short growth period in the spring and an even shorter one in the winter.

In fact the plant survives the summer season by losing its leaves and existing as a corm in a state of hibernation. After the summer, the plant again enters a period of vegetative growth with the emission of a tuft of leaves and the emergence of the floral axis wrapped in whitish sheaths.

Flowering takes place in the autumn, from the end of October to the middle of November. The flowers are made up of six mauve petals from which a scarlet stigma arises, which subdivides into three branches, each of which terminates in a tube. The stigma is connected to the ovary by a long style. The leaves, which grow up to 40 cm in length, are produced from September to May. It is in this same autumn–winter–spring period that root growth occurs, with reabsorption of the mother corm and production and growth of the daughter corms. Each newly formed corm, contained within the tunic of the corm which produced it, has one or two principal buds at its apex (from which new leaves, floral axis and one or two daughter corms are produced) and in the lower portion, four to five secondary buds, placed irregularly in a spir l form. The secondary buds produce a cauline axis and a tuft of leaves which draw nutrients through photosynthesis and grow. Corms derived from secondary buds are smaller (1/4–1/6) than the apical ones. Consequently, each mother corm produces two to three principal corms from apical buds and several corms from lateral buds. Saffron is a sterile species which exhibits effective vegetative reproduction.

Cultivated Area

L'Aquila saffron is cultivated in an atypical area considering the bio-ecological characteristics of the plant, almost at its ecological limits. In fact cultivation in Navelli takes place in a sub-mountainous area (plantations are between 650 and 1100 m above sea level), the highest area in the Mediterranean where saffron is cultivated, with an annual rainfall of about 700 mm, of which 40 mm falls in the summer.

In other saffron growing areas in the Mediterranean precipitation is lower; for example, in Greece at Kozani (Macedonia), annual rainfall is 560 mm, of which 25 to 40 mm fall in the summer; Spain (La Mancha and Castile) 250 to 500 mm, 20 to 30 mm in the summer; Sardinia (S. Gavino, Monreale) 300 to 600 mm, 20 to 40 mm in the summer.

Average annual temperatures in Navelli are also lower at 11.3°C, winter 2–5°C, summer 20–22°C; Kozani, about 12.5°C, winter 2–5°C, summer 23°C; La Mancha and Castile, 16–20°C, winter 5–7°C, summer 25°C; Sardinia, 16–20°C, winter 10°C, summer 25°C. The average summer temperature in Navelli never rises above 20–22°C, as compared to 25–30°C in the saffron growing areas in Spain and other parts of the Mediterranean. The xeric period in Navelli is limited to August; no xeric periods are recorded for other seasons. From December to January the average minimum temperature shows a negative value. Snow cover can last up to 30 days. Navelli saffron survives low winter temperatures without damage. However, heavy snowfall can damage the plants, especially if they are in flower: the flower freezes and decomposes and the corm splits and rots.

The environmental summer conditions (temperate-humid) in Navelli are largely responsible for cryptogamic attacks. In fact, the moisture and temperature values, especially those of the summer, create ideal conditions for the rapid development and spread of parasitic fungi (rot, decay, *Fusarium*). Massive attacks of parasitic fungi are recorded in the saffron plantations in Navelli when the spring is hot and rainy.

From observations, we have established that the critical temperature is the average March–April temperature of around 10–12°C (normal seasonal temperature being 6–9°C) accompanied by precipitation or dew. Under these climatic conditions, it is expedient to treat the soil or foliage with anti-fungal agents in order to save at least the daughter corms.

In Mediterranean areas characterized by a hot, dry summer climate, pluriannual cultivation of saffron is possible, particularly because the corms are not subject to devastating parasite attacks (hot, dry climates inhibit the reproduction and spread of parasitic fungi, mostly due to lack of water). Annual cultivation in the Navelli area consequently represents a strategy developed over the centuries so that the cultivation of saffron can continue in a sub-mountainous rainy environment at the ecological limits for this Mediterranean sub-desert plant.

The soil in the area under cultivation is a medium humus-clay, which guarantees good water storage, whereas the high sand content allows drainage and aeration. The active limestone content is good, organic substances high, phosphates low and potassium optimal.

PRINCIPAL AGRICULTURAL PRACTICES

Soil Preparation

Preceding the planting of corms, the soil is ploughed to a depth of about 30 cm and left to rest for a period from a few weeks to the whole winter. The cultivated area is divided up into plots of about 1000 square metres (20×40–50 m). A ridging hoe is used to prepare the bed; four parallel furrows, 2 by 2, are cut to a depth of about 10 cm for a length of 10–15 cm.

The corms are placed or lightly driven into place with the apex uppermost, generally in contact with one another (for more details see the section *Planting out*, below). They are then covered with the soil from the next furrow in line, to form a mound of about 10 cm in height. Four furrows make up a bed, locally called a patch. Each patch is about 80 cm wide, slightly raised to a height of 10–15 cm and about 50 cm long. The patches are separated from each other by a furrow, about 30 cm wide, which serves to give access for cultivation and above all acts as a drainage ditch.

Fertilizing

The soil in Navelli is fertilized with mature horse or cow manure (about 30 tons/ha). Contrary to cultivation practices in Spain and Greece, no mineral fertilizers whatsoever are used in Navelli. In Borgo Val di Taro (Parma), small saffron plantations have been planted and fertilized with both mineral fertilizers and manure.

Corm Harvesting

During June and July, the corms are dug up with a hoe, taken under cover and kept for a few weeks in hemp sacks. Before being replanted they are laid out on a canvas for individual examination and selection, based on the elimination of corms with cuts or marks and especially those with rot or parasites. The external tunics (2–3 layers) are then cleaned off, leaving only the interior tunic on each corm. The residual roots, in the form of a blackish-brown flattened disk, the residue of the previous year's corm, are then removed, as they can be the cause of fungal attacks.

Selection is based primarily on diameter and weight and only corms with a diameter larger than 2.5 cm are used; corms with too small a diameter are used as fodder (pigs, cattle). However, these should not be destroyed, as they can be planted in nursery beds, until their offspring reach the critical flowering size (diameter 2.5 cm) in subsequent years.

Planting Out

Corms in Navelli are not subjected to disinfection. Experience gained in Val di Taro (Parma) has shown that it is advisable to treat the corm, in order to inhibit the spread of disease, by immersion in a benomyl-based fungicide (5–10 mg per 1000 ml). In Spain and India, a solution consisting of 5% copper sulfate is used.

The planting period is the second fortnight in August (in Spain from 15–30 June; in Greece before the middle of September; in India from the middle of July to the end of August).

The preferred planting order in Navelli is four rows (two by two) per patch. In each row the corms are either in contact with each other or at a distance of 1–1.5 cm, with a planting depth of 8–10 cm. When the corms are planted to less than this depth the roots become large and fleshy (contractile roots), in which case the daughter corm does not grow, as the reserve material is stored in these roots and the mother corm is almost totally consumed by them. Contractile roots are also formed when the corms are disturbed during development by other factors (overcrowded planting, lacerations, etc.).

Tests have shown that the best yields – flower and corm production – are obtained by leaving a space of 2–3 cm between each corm in the furrow. The optimal quantity per hectare is 13–15 tons; that is about 600–700 thousand corms with an average weight of 20–22 g each (45–48 corms per kg). A hectolitre weight of 50–60 kg is equal to 2500 to 2700 corms. Therefore, about 1.3 tons of corms, that is 59–62 thousand pieces, are needed to plant an area of 1000 square metres (manual planting). Recently, experiments have been carried out using agricultural machinery similar to a modified potato planter (Galigani 1987, Galigani and Garbati Pegna, this volume).

Irrigation

Irrigation is not necessary in Navelli.

Weed Control

The saffron plantations in Navelli are infested with wild cereals which do not compete with saffron plants, because they are less developed during the period when the saffron plant is at anthesis (autumn) or at maximum growth (spring). As a result, no weed control is necessary. In fact the weeds are left to grow until the end of May, when they are cut together with the saffron leaves and used for cattle feed.

Production of hay is on the order of 60–80 kg/ha. The dry saffron leaves contain 12.12% nitrogenous substances and numerous mineral elements (about 7%) and have, therefore, good nutritive value. Their use as feed for cows and sheep results in increased milk production.

Flowering and Harvesting

Flowering occurs in autumn, about 40 days after planting, and lasts for about 3 weeks, from the middle of October to the 7th (10th) of November. A cold and snowy period, as well as late planting, can retard anthesis until after the middle of November. During anthesis, the highest concentration of flowers – over 60% of the plants in flower at the same time – occurs in the last 10 days of October. The Spanish call this period "the day of mantle", that is, the period during which the greatest expulsion of anthesis occurs, and the countryside becomes as though arrayed in a mantle of flowers.

The flowers are harvested manually. The picker moves between the patches picking the two rows to his left and the two rows to his right alternately. The flower is harvested by taking it between the thumb and the index finger of one hand and cutting it with the nail. The cut flowers are placed in a wicker basket to prevent them from being pressed together. The baskets are taken under cover and emptied onto a wooden table; "peeling" begins the same morning, i.e. the flowers are opened and the stigma is separated out.

It is impossible to mechanize this operation because flowering takes place contemporaneously with leaf growth and mechanized harvesting would involved cutting the leaves. As a result, the formation of daughter corms would not take place.

The flowers are picked early in the morning, while the flower is still closed, before the corolla opens. In this state the flower is quicker to pick and consequently easier and quicker to open for the removal of the stigma. Because the flowers have to be picked while they are still closed, working hours in the fields are limited to 2–3 hours in the morning. However, as many people as possible are needed during this phase in order to finish the work as soon as possible, because when the flowers open (opening occurs after sunrise when the soil heats up), according to local tradition, stigmas are considered to be of inferior quality. Stamens and anthers which are full of pollen are not picked; in Spain on the other hand, these are picked for their carotene and xanthophyll content.

Flower Production and Yield

Flower production in Navelli depends primarily on seasonal climatic conditions and on parasite attacks (rot, virus, etc.). A hectare of saffron plants yields 4–5 tons of fresh flowers; about 75 kg of fresh flowers are needed for 1 kg of fresh stigmas. The average weight of fresh stigmas from 100 flowers is 3.47 g and average dry weight is 0.69 g. The average weight of each flower is 0.3–0.5 g, each fresh stigma 30–40 mg, and each dry stigma 7–7.4 mg. There is a weight loss of 4/5 during the toasting process, and thus 1400 to 1500 flowers are needed to obtain 1 g of dry stigmas (the marketable product).

The average yield of the dry product per hectare is 10–16 kg. The saffron plantations in Navelli have the highest recorded production per hectare in the world. The yield of dried stigma filaments (kg/ha) elsewhere is 6–29 for Albacete (Spain), 4–7 for Krokos (Greece) and 1.8–6.8 for India.

Drying and Storing Methods

Separation of the stigma from the flower, called "stripping" or "peeling", is done by hand and carried out immediately after the flowers have been picked. The flowers (tepals) are opened and the stigma is cut with the fingers at the point where it divides into the three stigmatic branches, avoiding, as much as possible, any part of the yellowish style, as this lowers the quality of the product. The stigmas are laid on a sieve and placed about 20 cm above live oak-wood charcoal to dry. The sieve is connected by three ropes to a single support point, thus ensuring perfect roasting. Halfway through roasting the stigmas are turned over to ensure uniform drying. Roasting lasts for 15–20 min and drying is complete when the stigmas do not crumble and still possess a certain amount of elasticity when pressed between the fingers. Saffron dried over charcoal retains its purplish-red colour, its fragrance and its aroma. Results of trials carried out in electric drying ovens confirm that stigmas roasted in the traditional way over charcoal maintain their organoleptic qualities better (Zanzucchi 1987). During roasting, the stigmas lose 4/5 of their weight: 500 g of fresh stigmas yield only 100 g of dry stigmas. The final product retains 5–20% humidity. The dried stigmas are reduced to a powder by grinding in an electric coffee grinder. In a humid environment, saffron in filaments or powdered form is extremely hygroscopic and highly susceptible to fermentative processes, resulting in a change of colour and an unpleasant odour. It is therefore kept in well-sealed, coloured-glass jars (without rubber stoppers) or in canvas bags, and stored in a dry place.

Crop Rotation

The cultivation of saffron is never carried out on the same plot within at least 10 years of the previous saffron crop. In some cases where this custom has not been respected, a decrease in production was observed, with an increase in the number of weeds attributable to the previous saffron crop. Saffron crops are rotated with lucerne and wheat.

Pests and Diseases

The saffron plantations in Navelli are subject to adverse climatic conditions which cause damage to the part of the plant which is above ground, and to attacks by rodents (moles and mice) which damage the corms. The most ruinous diseases, however, are of fungal origin.

In the Navelli area, a mycosis produced by *Penicillium cyclopium* is particularly prevalent, causing mauve-coloured rot as a result of insect attacks. It is at its most virulent during the hot, humid season. In the past 15 years, the plantations have been attacked by an alarming disease which causes abnormal leaf growth (up to 50 cm) and over-development of the floral sheath. The plant becomes thin and white as the sheath forms a sleeve, which prevents the leaves and flowers from emerging, even though they are perfectly formed. The corm cells deliquesce and the corm gradually dissolves. Plants attacked by this disease are called "little candles" by the growers. The pathogenic agent appears to be *Fusarium*, a microscopic fungus producing gibberellin, which causes abnormal growth of the leaves and sheaths. This parasitosis is also more prevalent during the hot rainy season, reducing flowering by 10 to 30%. Where plants are left in the soil for two years, the disease reaches 45%. This is yet another reason why pluriannual cultivation is not feasible in Navelli.

USES

At present saffron is used mainly in the liqueur industry (aperitifs, bitter, vermouth) and in the confectionery industry, for the colouring and flavouring qualities of its active components. In the food industry and in cooking it is used as a colouring for pasta and cheese, and in the preparation of regional specialities (risotto alla milanese, paella valencians, etc.). In the Navelli area it is used in cooking, sold by herbalists and grocers, and used in the preparation of local liqueur.

QUANTITY, ACREAGE AND PRICES

Over the centuries and until 40 years ago, Navelli saffron was cultivated over a remarkably large surface area. At the beginning of the 20th century more than 450 ha were under cultivation, producing 4.6 tons, and in some years the cultivation area exceeded 1000 ha, extending into other Abruzzo valleys (Sulmona, Marsica). For inland Abruzzo, saffron was an authentic and economic source of wealth; for example, the prices quoted for saffron in the 15th and 16th centuries were higher than per equal weight of silver, and saffron fields were therefore considered to be more remunerative than silver mines. Even in the 20th century, up to the 1960s, the area under cultivation was on the order of 180–200 ha, producing 2 tons.

In the past 30 years, however, the area under cultivation at Navelli has been heavily reduced, due to socioeconomic factors (population shift from the countryside, an increase in service industries, etc.) reaching an all-time low in 1976 of 3.5 ha, and producing only 20 kg of saffron. In the past few years there has been a revival in cultivation, with production reaching 40 kg of dry saffron from 1985–1987 (cultivation area 6 ha), and in 1988–1995, production exceeded 80 kg from an area of around 8–9 ha.

Both the surface cultivated and the production of dry stigmas are modest as compared to Spain (2864 ha; 12.9 tons in 1983) and Greece (860 ha; 3.7 tons in 1988). However, even though the quantity of saffron produced in Navelli is small, it merits the highest consideration, because of the bioagronomic characteristics of its germplasm, its outstanding organoleptic qualities for cooking and food preparation, and the fact that it represents a source of income for the few remaining saffron growers at Piano di Navelli. Throughout history, Navelli saffron has overcome crises in production of a far more serious nature than any known today (in 1646, during the Spanish domination, production almost ceased and only three pounds were produced vs. 12 thousand, 200 years earlier. This was because a decree issued by the viceroy gave foreign buyers the exclusive right to set prices).

Techniques of *in vitro* culture and other innovative methods presently under study (elimination of sterility, hybridization) offer a realistic possibility of the revitalization of saffron cultivation in Navelli. The main purpose of these studies is to make sufficient material available for planting, in particular in view of the interest shown in saffron cultivation by the young, who find it impossible at present to cultivate the plant due to lack of corms.

L'AQUILA SAFFRON: A TYPICAL ITALIAN PRODUCT

Measurements and statistical biometric comparisons have been made between L'Aquila saffron and that of Krokos (Greece), Pozo Hondo (Spain) and S. Gavino (Cagliari, Sardinia), and an F test (variance analysis) was carried out. Taking into account the significant differences derived from the statistical analysis of the principal characteristics, the high annual yield of stigmas per hectare, their strong colouring power and high safranin content, saffron plants from Navelli (L'Aquila) differ from those cultivated in Spain and Greece. They represent a typical Abruzzo and Italian cultivar which is characterized by the weight of the corm (22.9 g), its diameter (3.23 cm), the annual spice yield (10–12 to 16 kg dry stigmas/ha), and their high safranal content (4%).

This is why we have classified Navelli saffron as *Crocus sativus* L. cultivar Piano di Navelli – L'Aquila, in honour of the city and district which for five centuries has been the home of saffron cultivation in Italy. This classification guarantees a market for Navelli saffron. Recently (Tammaro 1994), the saffron from Navelli (L'Aquila) was recognized as a typical regional product of the European Community (EC) and a logo for the product's preservation is being designed.

The above cultivation, in fact, fits the criteria of the European Economic Community (EEC) n. 2081 and 2082 (July 14 1992) rules, i.e.:

(a) strong presence in local historic culture (tradition, local uses and holidays);
(b) a typical geographical localization of production;
(c) high organoleptic quality of the product;
(d) production techniques belonging to an exclusive typology.

At present, saffron in Italy shows little economic value due to the scarce harvest. On the other hand, it is of great scientific importance since it represents a typical Italian cultivar, which has been selected for over the past 500 years. Moreover, saffron is remarkable for the rural tradition of Southern Italy, especially the Abruzzo Region, which has based a certain amount of its economy on this plant for the past few centuries.

REFERENCES

Galigani, P. (1987) La meccanizzazione delle colture di salvia, lavanda, zafferano e genziana. In A. Bezzi (ed.), *Atti Convegno sulla coltivazione delle piante officinali, 9–10 ottobre*, Istituto Sperimentale per l'Assestamento Forestale e per l'Alpicoltura, Villazzano (Trento), pp. 221–234.

Picci, V. (1987) Sintesi sulle esperienze di coltivazione di *Crocus sativus* L. in Italia. In A. Bezzi (ed.), *Atti Convegno sulla coltivazione delle piante officinali, 9–10 ottobre*, Istituto Sperimentale per l'Assestamento Forestale e per l'Alpicoltura, Villazzano (Trento), pp. 119–157.

Tammaro, F. (1990) *Crocus sativus* L. cv. Piano di Navelli (L'Aquila saffron): environment, cultivation, morphometric characteristics, active principles, uses. In F. Tammaro and L. Marra (eds.), *Proceedings of the International Conference on saffron (Crocus sativus L.), L'Aquila 27–29 October 1989*, Università degli Studi L'Aquila and Accademia Italiana della Cucina, L'Aquila, pp. 47–98.

Tammaro, F. (1994) Lo zafferano di Navelli (*Crocus sativus* L.). *Programma di Iniziativa Comunitaria LEADER 1* (U.E.), L'Aquila (Italy), 1–44.

Zanzucchi, C. (1987) La ricerca dal Consorzio Comunalie parmensi sulla zafferano (*Crocus sativus* L.). In A. Bezzi (ed.), *Atti Convegno sulla coltivazione delle piante officinali, 9–10 ottobre*, Istituto Sperimentale per l'Assestamento Forestale e per l'Alpicoltura, Villazzano (Trento), pp. 347–395.

ENDNOTES

1. According to Galigani and Garbati Pegna (this volume), saffron was recently reintroduced to San Giminiano (Tuscany).

6. SAFFRON CULTIVATION IN AZERBAIJAN

N.SH. AZIZBEKOVA[1] and E.L. MILYAEVA[2]

[1]*Plant Science Department, University of British Columbia,*
#344 – 2357 Main Hall, Vancouver, B.C., V6T 1Z4, Canada
[2]*Timiryazev Institute of Plant Physiology, Russian Academy of Sciences,*
Botanicheskaya Street 35, Moscow, 127276 Russia

ABSTRACT *Crocus sativus* (saffron) has been cultivated in Azerbaijan for centuries. Its distribution, adaptation and cultural practices there are discussed. Saffron ontogenesis, with special reference to the morphological state of the cells and nuclei in the stem apex, is detailed. Stem-apex development in saffron followed a seasonal pattern: (1) formation of stem-apex of the daughter corm in November, (2) slow development of stem-apex of the daughter corm coinciding with intensive plant vegetative growth of the maternal corm in December–February, (3) transition of stem-apex of the daughter corm to generative development when the vegetative organs begin to dry up in March, and (4) differentiation of generative organs when the corms are underground and other vegetative organs are almost fully absent in June–August. The treatment of saffron corms with gibberellin promoted the formation of flower buds from undifferentiated meristems, thereby increasing stigma yield. The best results were obtained when the corms were soaked in gibberellin in July.

DISTRIBUTION AND ADAPTATION

Azerbaijan is one of the oldest centres of saffron (*Crocus sativus*) in the world. According to written testimonies, saffron was cultivated in some regions of Azerbaijan more than a thousand years ago (Askerov 1934), the practice having been introduced from Asia Minor and Persia. However, a more precise date for the appearance of saffron cultivation in Azerbaijan has not been determined. Saffron escapes have been found in the foothills of the main Caucasian mountain range in the Cubinsky, Shamakhinsky and Gueokchaiksy regions in Azerbaijan (Grossheim 1940).

At the time of the Roman Empire, saffron from Asia Minor and central Asia was introduced to Spain, and from there to the south of France. After the Crusades, its cultivation began in Germany, Austria and Moravia. However, saffron cultivated in northern regions is characterized by its lower quality relative to that cultivated in areas similar to its native land. The best saffron, possessing the most powerful aroma, is cultivated in Spain, Iran and Azerbaijan.

In Azerbaijan, saffron is cultivated on the sandy Apsheron peninsula near the city of Baku. In the past, this culture existed in Mashtagy, Bylgya, Kurdakhany, Nardaran and other settlements. There the choicest saffron was grown, with quality characteristics matching those of Persian and Spanish saffron. Before 1917, the areas occupied by cultivation amounted to 150 ha, and saffron export afforded Azerbaijan a significant place in the international market (Gurvich and Zadulina 1939). Under Soviet rule the areas devoted to saffron cultivation in Azerbaijan grew 4.5-fold, and in the Bylgya settlement, near Baku, a specialized saffron state farm was established.

The climate in the Apsheron peninsula is particularly favourable for saffron, being subtropical, characterized by dry and warm autumn months: the maximum, minimum and average annual air temperatures are 33.2°C, – 5.9°C, and 14.4°C, respectively.

Saffron in Azerbaijan flowers in autumn – in October–November. The most influential factor in its successful cultivation is atmospheric humidity, particularly during flowering. The average precipitation on the Apsheron peninsula is up to 223 mm per annum, with 72% relative humidity, an atmospheric pressure of 648 mm Hg, and an average annual temperature of the sea-water surrounding the peninsula of 15.2°C. During the saffron flowering period, the weather is usually warm and dry: in August precipitation averages 7 mm, relative humidity is 64%, atmospheric pressure is 607 mm Hg, and the sea-water temperature is 25.8°C; in September, average rainfall increases two-fold, relative humidity is 68%, atmospheric pressure is 643 mm Hg, and the sea-water temperature is 22.9°C; in October, rainfall is higher still (27 mm), relative humidity is 75%, atmospheric pressure is 678 mm Hg, and the sea-water temperature is 18.3°C (Kadymov 1940).

Saffron is quite fastidious in its soil requirements. The soil needs to be light and friable, with high nutrient content. On the Apsheron peninsula there are chestnut and shined-chestnut in the foothills and loamy sandy soil near Baku. Saffron can be cultivated practically anywhere in the peninsula, except in areas with gravele-clayey soil, where poor drainage may lead to corm decay. Such conditions may occur, for instance, in Mashtagy saffron plantations where middle loam, containing 1:1.8:1.25 clayey particles/sandy-dust/sand, dominates. The amount of humus in this light soil is not high – from 0.5% (at a depth of 0–15 cm) (Kylany 1979). As a result, organic fertilizers (manure) are added to fields intended for saffron cultivation. However, insufficiently fermented fresh manure causes corm decay. Fresh manure is therefore added to one of the grain cultures, wheat or barley, preceding saffron.

A unique aspect of saffron cultivation on the Apsheron peninsula is the absence of precise intervals between crops in crop rotation: as a rule, saffron is grown in the same area for 3 to 5 years. A field intended for corm planting is carefully ploughed in the autumn to a depth of 50 cm, cleared of weeds, and treated with manure; again, in the summer, it is cleared of weeds, harrowed and treated once more with manure, at which point the corms can be planted.

When saffron is cultivated in lowland areas, wind-barriers are necessary, usually consisting of camel's-thorn (*Alhagi camelorum* L.). Saffron plantations on hilly terrain need to be situated on the southern, eastern or western slopes, which are protected from the winds.

The corms are dug up from the old growing area in the summer months, from July to August. Before planting, corms are kept in heaps or split into layers, stored, and cleaned; sick and decaying specimens are discarded. Corms grown under optimal conditions have a flattened-globous shape, an average weight of 8–10 g, and a shiny-cream colour; they are covered with golden-rose tunics, above which there are rough, stringy, dark-brown external tunics.

The corms are planted in 8-m long rows. The furrows are 30 cm deep, 60 cm apart, and the distance between corms is 10 cm. Post-planting consists of weeding and crumbling.

ORGANOGENESIS

Saffron has a unique life cycle. It belongs to the group of ephemeral geophytes: flowering occurs in autumn, at the end of October to the beginning of November (Azizbekova and Milyaeva 1978). Upon completion of flowering, a small daughter corm is formed at the base of the main maternal-corm shoot (Figure 6.1). More daughter corms are formed at the bases of side shoots sprouting on the maternal corm. During this period, intensive growth of leaves and roots occurs, attaining their maximal development during the winter months (December–February). The maternal corm gradually dries up and dies concomitant with the intensive growth and development of the daughter corms during this period. The leaves, root system and maternal corm begin to dry up gradually at the end of March. The concealed, underground period in the life of corms occurs from May through August, inasmuch as leaves and roots are totally lacking. The central bud of the corm is induced into growth by the increased moisture and reduced temperatures in August–September, and root primordia appear at the base of the (new) maternal corm.

Thus, the ontogeny of the saffron crocus can be divided into three major periods: flowering, vegetative growth, and summer "dormancy". In a previous study we showed

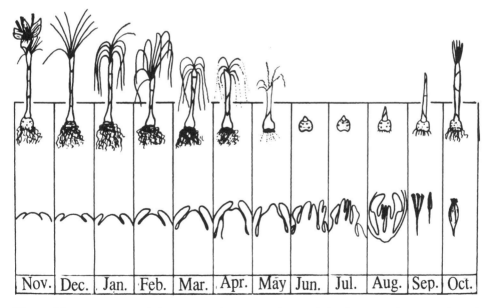

Figure 6.1 Ontogenesis of *Crocus sativus* plants (upper line) and the changes in the stem apices of their buds (lower line).

that changes in the physiological state of the stem-apex meristem occur in conjunction with these periods (Azizbekova 1978). During the first month of daughter-corm formation, in November, the stem-apex meristem is small and slightly convex. The cells and nuclei are at their smallest relative to the other months (Figure 6.2a, Table 6.1) (Milyaeva and Azizbekova 1978). Staining with Schif's reagent gives the nuclei a dense, brightly stained appearance (Figure 6.3a), The mitotic index at this time is equal to 4.8 (Figure 6.4a). Figure 6.5 represents histograms of nuclear distribution with a DNA content of 2C–4C. In November, 75% of the nuclei contain 30 to 50 arbitrary units of DNA, corresponding to the 2C state. This means that a high percentage of nuclei are in the G_1 phase of the cell cycle during this period.

Insignificant changes are discernible on longitudinal sections of apices during the next ontogenetic period (from January through February): the apex changes its configuration slightly, leaf primordia protrude (Figures 6.2b and 6.2c), and the mitotic index increases slightly (Figure 6.4a). The dimensions of the cells and nuclei undergo virtually no changes (Figures 6.3b and 6.3c, Table 6.1). The histogram (Figures 6.5b and 6.5c) reveals a slight increase in the number of nuclei with intermediate DNA content (49–50 arbitrary units) during this period. Thus we assume that some nuclei pass from the G_1 phase to the S phase of the cell cycle.

A sharp change in all of the nuclei studied is observed in March. Apex size increases abruptly, where the intensive establishment and growth of leaf primordia

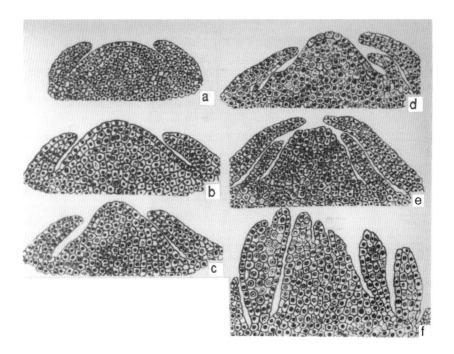

Figure 6.2 Longitudinal sections through saffron stem apices at different developmental stages: (a) in November; (b) in January; (c) in February; (d) in March; (e) in May; and (f) in June. Magnification × 200.

Table 6.1 Changes in nuclear dimensions during the course of stem-apex differentiation in *Crocus sativus* (in arbitrary units)

Month	Number of measured nuclei	χ	$\pm\delta$		t	
November–February	100	3.48	0.6	} 13.6		
March	100	5.13	1.1	} 6.1	} 28.6	
April–June	100	5.91	0.8			

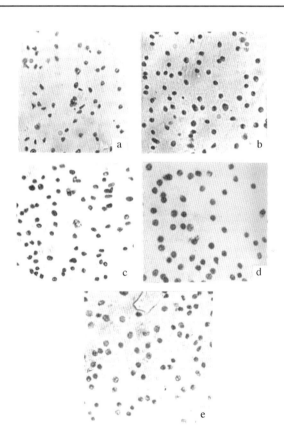

Figure 6.3 Micrographs of nuclei in the cells of saffron stem apices at different developmental stages: (a) in November; (b) in January; (c) in February; (d) in March; and (e) in May. Magnification × 900.

proceeds, and the laying down of generative primordia begins (Figure 6.4a). Dimensions of cell (Figure 6.2a) and nuclei (Table 6.1, Figure 6.3b) also increase. As can be seen from the histogram (Figure 6.5d), a large percentage of nuclei pass over into the G_2 phase of the cell cycle.

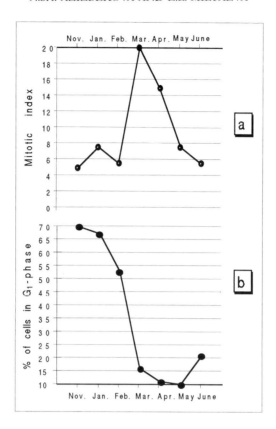

Figure 6.4 The mitotic index (a) and the distribution of nuclei in the G_1 phase of the cell cycle (b) in saffron stem apices at different stages of development.

Apex differentiation progresses further in May–July, during which time intensive laying down of generative organs occurs (Figures 6.2e and 6.2f). The mitotic index decreases during this period, and stabilizes at a constant level at the beginning of June (Figure 6.4a). The number of cells in the G_1 phase declines relative to March (compare Figures 6.5d–6.5g), whereas the number of cells in the G_2 phase rises, then remains virtually unchanged from May through June, matching the constant mitotic-index value from May to June. Nuclear dimensions increase still further, as compared to March, during this period of intensive creation of generative organs (Figure 6.3a, Table 6.1).

The stem apices undergo continuous structural and cytophysiological changes during ontogenesis of *Crocus sativus*. During the period of stem–meristem establishment, most of the cells and nuclei are small, brightly staining and in the G_1 phase. The remaining nuclei are in the S and G_2 phases. Cellular and nuclear dimensions increase during the period of generative organ formation from April through June, whereas mitotic activity decreases slightly, and most nuclei pass over into the G_2 phase. Therefore, the flowering of the maternal corm coincides with the formation of daughter corm(s) during November; vegetative growth of the leaves and roots of the maternal corm coincides with the slow development of the stem-apex of the

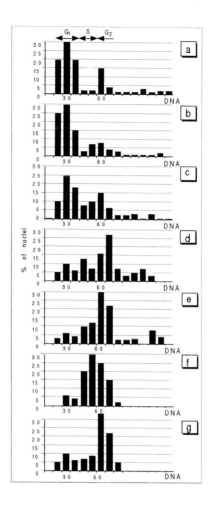

Figure 6.5 Histograms of the distribution of nuclei with a DNA content of 2C–4C (given in arbitrary units) during stem-apex development: (a) in November; (b) in January; (c) in February; (d) in March; (e) in May; (f) in June; and (g) in July.

daughter corm (December–February). On the other hand, the gradual senescence of leaves, roots and the maternal corm coincides with stem-apex transition to generative development (March). Intensive differentiation of flower organs occurs when daughter corms are "dormant" (June–August).

THE BENEFITS OF GROWTH REGULATORS

Exogenuously applied growth regulators (gibberellin and kinetin) have been investigated in *Crocus sativus* in relation to floral development. One of the most important

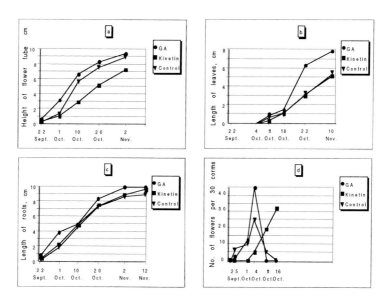

Figure 6.6 Influence of gibberellin and kinetin on growth and flower number of saffron: (a) length of flower tube, (b) length of leaves, (c) length of roots, and (d) number of flowers.

Figure 6.7 Flowering saffron plants in an experiment conducted in Bylgya. Plants after treating corms with gibberellin (left) and control (right).

factors in such study is the precise physiological stage of plant development at the time of growth-regulator application.

Saffron corms were soaked in a solution of gibberellin in February (before the apex had differentiated generative tissue), in March (during the transition from vegetative to reproductive) and in June (when the floral organs are present) (Azizbekova *et al.* 1978). The most positive results stemmed from a June–July treatment of gibberellin+kinetin, when corms were dormant and underground (Figure 6.6). Treating dry saffron corms with growth regulators in June–July promoted the formation of additional flower buds from undifferentiated meristem (Figure 6.7). This led to the accelerated formation of more flowers, which in turn increased saffron yield.

REFERENCES

Askerov, A. (1934) [Saffron]. Azerneshir Publishing, Baku, 23 pages (in Russian).

Azizbekova, N.Sh. (1978) [Cytphysiological changes in the course of stem apices development of saffron crocus (*Crocus sativus* L.).] Ph.D. thesis, Institute of Plant Physiology, Moscow (in Russian).

Azizbekova, N.Sh. and Milyaeva, E.L. (1978) [Ontogenesis of saffron (*Crocus sativus* L.) and the changes in stem apices. *Ontogenesis*, 9(3), 309–314 (in Russian).[1]

Azizbekova, N.Sh. and Milyaeva, E.L.(1979) Ontogenesis of saffron crocus (*crocus sativus*) and changes in stem apices. *Soviet Journal of Development Biology*, 9, 266–271.

Azizbekova, N.Sh. and Milyaeva, E.L. (in press) Saffron cultivation in Azerbaijan (this volume).

Azizbekova, N.Sh., Milyaeva, E.L., Lobova, N.V. and Chailakhyan, M.Kh. (1978) Effects of gibberellin and kinetin on formation of flower organs in saffron crocus. *Soviet Plant Physiology*, 25(3, part 2), 471–476.

Milyaeva, E.L. and Azizbekova, N.Sh. (1978) Cytophysiological changes in the course of development of stem apices of saffron crocus. *Soviet Plant Physiology*, **25** (2, part1), 227–233.

Azizbekova, N.Sh., Milyaeva, E.L., Lobova, N.V. and Chailakyan, M.Kh. (1978) [Effect of gibberellin and kinetin on formation of flower organs in saffron crocus. *Russian Plant Physiology*, **25**, 471–476 (in Russian).

Grossheim, A. (1940) [*Flora of Caucasus*] Baku, Vol. II, 200–203. (in Russian).

Gurvich, N.A. and Zadulina, V.I. (1939) [Saffron]. Azerbaijan Branch of the Academy of Sciences Publishing, Baku, 130 pages (in Russian).

Kadymov, D.R. (1940) [Directions for agriculture of saffron in Apsheron]. Azerneshir Agricultural Department Publishing, Baku, 8 pages (in Russian).

Kylany, A.N. (1979) [Growing of saffron, directions for agriculture]. Azerneshir Publishing, Baku, 24 pages (in Azerbaijani).

Milyaeva, E.L. and Azizbekova, N.Sh. (1978) [Cytophysiological changes in the course of stem apices development of saffron crocus]. *Russian Journal of Plant Physiology*, **15**(2), 289–295 (in Russian).

---

[1] Some of the Russian references cited here have been translated into English.

7. SAFFRON CULTIVATION IN GREECE

APOSTOLOS H. GOLIARIS

*Department of Aromatic and Medicinal Plants,
Agricultural Research Centre of Macedonia — Thrace,
570 01 Thermi — Thessaloniki, Greece*

ABSTRACT   *Crocus sativus* L. is a perennial plant that propagates by corms. Today, systematic saffron cultivation in Greece is confined to Kozani county, western Macedonia, and is controlled by the Saffron Growers' Cooperative. The crop may be kept economically profitable until the seventh year. Sample analysis from many fields, over a number of years, has proved the excellent quality of Greek saffron.

INTRODUCTION

The ancient Greek word *krokos* (saffron) refers, in its broadest sense, to the plant, the flower, the dyeing substance, the aromatic oil and the pharmaceutical herb. Etymologically the word *krokos* comes from the Greek word *kroke*, used to designate the yarn woven with a shuttle in the warp of a loom. A famous fresco in the Minoan palace of Knossos, Crete, dated from 1600 BC and known as the "saffron gatherer" depicts a blue monkey picking saffron flowers. Hippocrates (470–377 BC), Aesculapius (525–456 BC), Theophrastus (372–287 BC), Dioscorides (first century AD) and Galen (129–201 AD) quote the word *krokos* with reference to the pharmaceutical herb. Sophocles (496–406 BC), the classic Greek poet and dramatist, quotes the word *krokos* "golden dawn *krokos*" in his drama "Oedipus on Kolonos" to denote the plant. In the hymn to Demeter in his Iliad, Homer (10th–9th century BC) speaks of the flowers of *krokos*. Aeschylus (529–456 BC), in his drama "Agamemnon", reports that Darius' sandals were dyed with *krokos* (saffron). Aristophanes (445–385 BC), in Thesmiotes, reports that the tunics worn by Dionysus and his followers during the Dionysian mysteries were dyed with *krokos* (saffron). The word *krokos* also appears once in the Greek translation of the Old Testament in Songs of Solomon 4.13–14.[1]

There are a number of theories concerning the origin and spread of the species. Some scientists support the view that the saffron plant is native to the Orient. Others believe that the species originated in Greece, where it was domesticated and cultivated for the first time during the Minoan period. This theory is strengthened by "The saffron gatherer" fresco of that period found in the palace of Knossos on

---

[1] And saffron in the King James translation: "Thy plants are an orchard of pomegranates, with pleasant fruits; camphire, with spikenard. Spikenard and saffron [*karkom* in Hebrew]; calamus and cinnamon, with all trees of frankincense; myrrh and aloes, with all the chief spice."

Crete. Subsequently *Crocus* cultivation spread in the Near and Middle East, probably at the time of Alexander the Great, king of Macedonia, in the 4th century BC.[2]

Today, systematic saffron cultivation in Greece is confined to only two villages (*Krokos* and Karyditsa) in Kozani county, in western Macedonia. In the past it was also grown on the islands of Crete, Thera, Anafi, Delos, Syros, Tenos, Aegina, Mykonos, Andros and Corfu.

Eighteen *Crocus* species grow in the different geographical regions of the Greek islands and mainland:

1. *Crocus chrysanthus* Herb., indigenous to hilly areas throughout Greece.
2. *C. olivierii* Day, indigenous to mountainous areas all over Greece.
3. *C. biflorus* Mill., indigenous to northern Greece and the Ionian Islands.
4. *C. crewii* Hook., indigenous to mountainous areas all over Greece.
5. *C. veluchensis* Herb., indigenous to mountainous areas of continental Greece.
6. *C. sieberi* Day, endemic plant in the mountainous areas of Crete.
7. *C. nivalis* Bory & Chaub., indigenous to the alpine areas of continental Greece.
8. *C. atticus* Orph., indigenous to sub-alpine areas all over Greece.
9. *C. pulchellus* Herb., indigenous to the area from northwestern Greece down to Thessaly.
10. *C. tournefortii* Gay, or *C. boryi* var. tournefortii Baker, *C. orphanidis* Hook., indigenous to the Cyclades, predominantly on Syros, Tenos, Mykonos and Delos.
11. *C. veneris* Tappein., indigenous as an endemic plant in the island of Crete.
12. *C. boryi* Cay., indigenous to Thessaly, Peloponnese and Crete.
13. *C. levigatus* Ch. & Bory, indigenous to Thessaly, continental Greece (Sterea Hellas), Peloponnese and Crete.
14. *C. sativus* L., exists only as a cultivated plant in Greece.
15. *C. cartwrightianus* Herb., indigenous to the low-fertility areas of Attica, the Aegean islands and Crete.
16. *C. hadriaticus* Herb., indigenous to mountainous areas all over Greece.
17. *C. peloponnesiacus* Orph., indigenous endemic plant on Malevon Mt., Laconia county.
18. *C. cancellatus* Herb., indigenous all over Greece. The corms of said species are edible after cooking or seasoning.

Of these eighteen species, the fertile *C. cartwrightianus* is considered to be the progenitor of the sterile *C. sativus* (Mathew, this volume).

CULTIVATION

Botany

Crocus sativus L. is a perennial plant having a depressed globule-shaped underground corm, 3–5 cm in diameter. The leaves are narrow, grass-like, 30–50 cm long. The flowers, one to four per corm, open before leaf emergence, and consist of six violet petals expanding outwards at the top. The pistil is made up of a bulbous ovary from

---

[2] Or even earlier, see Negbi's article in this volume.

which a slender style arises which is pale yellow and divides into a brilliant orange-red, three-lobed stigma, 3–5 cm long. There are three stamens per flower, with two-lobed anthers.

Climate

Saffron begins its growth in autumn, retains its leaves in winter, and enters a dormant state by the end of spring, so as to escape the high summer temperatures. A mild subtropical climate is considered most suitable for saffron cultivation.

The regions in which saffron is grown in Greece are characterized by a specific microclimate: annual precipitation exceeding 500 mm, 6–7°C average minimum temperature and 13.5–19°C average maximum temperature during October and November. The crop endures drought, but at certain stages of its growth water is indispensable. These critical times, when rain or irrigation is necessary, include March and April, when the corms grow, and September, for quantitative and qualitative improvement of the crop.

Soil

Saffron grows in a wide range of soils, but thrives best in deep, well-drained clay-calcareous soils that have a fairly loose texture and permit easy root penetration. The soils need not be rich in nutrients. However, low- and high-pH calcareous soils, as well as poorly drained ones, are unsuitable. An analysis of four typical soils in which saffron is cultivated in Greece is shown in Table 7.1.

Propagation – Lifting up of the corms

Saffron is propagated by corms. Each mother corm produces three or four new corms in the subsequent (second) year, while the mother plant itself decays. In the third year, 1–6 new corms are produced from each mother corm of the previous year, which also then decays. In the fourth year corm production declines, so only one or no corms are produced from a mother corm, which itself decays. This continues until the fifth and sixth years. Thus, in the position occupied by the initial corm in a new plantation (first year), one finds 3–4 corms in the second year, 20–22 corms in the third year and 18–20 corms from the fourth year onwards. The new corms

Table 7.1 Analysis of four typical soils supporting saffron cultivation in Greece

No.	Depth (cm)	Texture Class	Free $CaCO_3$	pH Saturated Paste	Organic Matter %	Available Nutrients		
						P (ppm) (Olsen)	K_2O (mg/100g) (Dirks)	B (ppm)
1	0–30	SCL	+++	7.30	2.65	7.62	7.50	0.07
2	0–30	CL	+++	7.40	1.40	4.48	7.25	
3	0–30	CL	++	7.45	2.35	12.60	7.00	0.25
4	0–30	CL	++	7.45	1.24	3.40	6.20	

SCL = sandy clay loam, CL = clay loam, +++ = more, ++ = medium

begin to form after the November blossom and complete their development before the foliage dries out in May.

The plantation may be kept economically profitable until the seventh year by exploiting the flowers for a number of years. The lifting up and harvesting of corms to be used as propagation material takes place after leaf drop, from June to September, preferably from old (5 to 7 year) plantations, which can produce nearly 6–7 tons of well-formed healthy corms per ha. The corms are stored in a cool, dry place. Within a maximum of 2 months they can be transplanted in the field to establish a new plantation. For a hectare of new plantation, 2–3 tons are needed. A well-developed corm should be 22–25 mm diameter and 35–40 mm high on average. Before transplanting, the corms must be dipped into a fungicide solution (usually PCWB 75 W.P.) for 5 min. The suggested application rate for PCWB 75 W.P. is 150 g active ingredient per 100 kg of water.

Field Preparation Before Transplanting

A new plantation can be established between May and September. Before transplanting, the land should be well prepared to a depth of 30–35 cm by ploughing two or three times, depending on the prevailing climatic and soil conditions. The first ploughing is performed in July and the second in August. Before the second ploughing, 20–30 tons per ha of well-fermented animal manure are spread over the land surface, to be incorporated into the soil by the subsequent ploughing. The third ploughing is performed 8–10 days before transplanting for fine soil preparation and incorporation of mineral fertilizer. After the last ploughing, drainage troughs are formed every 10–12 m to ensure good drainage of the field in case of heavy rainfall.

Corm Planting

The corms used to be planted in furrows formed with a plough. The workers placed the corms upright in the rows, 11–13 cm apart along the row at a depth of 15–17 cm.

The corms were covered with the soil turned over by the plough as the next furrow was formed. The distance between the rows is 20–25 cm. Therefore 230,000–250,000 corms per ha are needed to obtain a good plantation. Corm planting is followed by a light harrowing. Nowadays, the transplantation is carried out by machine, which was sophisticated by the growers themselves. The best period to establish a new plantation is June. After that, no more cultivation is needed until September, when superficial chiselling can be performed to a depth of 6–8 cm.

Cultivation in the Old Plantation

When the plants in the old plantation (second to sixth years) begin to dry out in May, all the weeds are cut and removed from the field. The soil is then cultivated to a depth of 10 cm. The first cultivation consists of chiselling in early June, and this is then repeated in July and September.

Fertilization

At the last chiselling in September, 40 units of N, 30 units of P_2O_5 and 40 units of K_2O per ha are applied and incorporated. Some growers apply another portion of 30 units of N the following March as a surface dressing in NO_3^- form. In some instances chlorotic symptoms are observed, and these are attributed to Fe or Mn deficiencies. To confront or alleviate Fe deficiency, chlorotic plants are watered with an organic Fe solution (Sequestrene 138, Fe), at a rate of 30 g of organic Fe per 10 m^2 of soil, dissolved in a small quantity of water. When Mn is deficient, soil applications of 200 kg per ha $MnSO_4$ are used, or the plants are sprayed with an aqueous solution of 1‰ $MnSO_4$.

Weed Control

Many weeds compete with the crop. Most common are *Anagalis arvensis* L., *Amaranthus blitum* L., *Avena fatua* L., *Capsella bursa pastoris* L., *Cichorium intybus* Jacq., *Fumaria officinalis* L., *Papaver rhoeas* L., *Sinapis arvensis* L., and *Sonchus oleraceous* L.

The best weed-control method consists of hand-weeding, hoeing. These are the most effective and environmentally friendly ways, but also the most expensive. Another way is the light chisseling. The work begins after flower-picking in November and lasts till April. Over the past few decades, scientists have explored and experimented with the use of more and more herbicides for weed control. According to our trials, the best control is achieved with the herbicides Simazine (Gesatop 50%) and atrazine (Gesaprim 50%) at a rate of 10 kg per ha.

Diseases and Pests

The most serious fungal disease of saffron is *Rhizoctonia crocorum* (PERS) D.C. which causes corm decay. There are several ways to control this fungus:
(a) removal and burning of the infected plants, (b) in fields heavily infested with this fungus, a 5-year crop rotation is advised, and (c) watering the root system of diseased plants when the first symptoms became apparent with a curative solution of the fungicide P.C.N.B. W.P. (Brassicol) at a rate of 1.5–3 g active ingredient per m^2.

Other pests that cause serious damage to saffron plantations are rats, which eat the corms, and moles, which destroy them. Rats can be effectively controlled using poisonous baits and moles by using a smoking gas apparatus or poisonous-gas-releasing tablets placed at the entrance to their tunnel. Special handmade guns have also proven satisfactory against these pests.

Flower Picking

Saffron flowers are ephemeral. If they are exposed for too long to sunny, windy or rainy weather, their stigmas and styles lose their colour and quality and their perfume deteriorates. Flowers must therefore be picked daily during peak flowering and every other day at the beginning and end of the flowering period. The flowering period starts around the beginning of October and lasts for about 30–40 days, depending on the prevailing weather conditions. Peak full-flowering coincides with the second decade of October. The flowering period of each plant may last for up to 15 days (Figures 7.1–7.3).

Flowers are picked by hand, from sunrise to sunset (Figures 7.4 and 7.5). The flower is cut at the base of the petals with a slight twisting movement or with the fingernail. The cut flowers are collected in baskets. The largest saffron yields are

Figure 7.1 Fully grown saffron plants from a commercial crop in Krokos, Greece.

Figure 7.2 Saffron crop at the fully flowering stage in Kozani county, Greece.

SAFFRON CULTIVATION IN GREECE

Figure 7.3 Flowering saffron crops. Styles and stigmas are quite distinct.

obtained from third- and fourth-year plants. A product with excellent quality characteristics is reaped under temperatures ranging from 13°C to 19°C and a relative humidity of 60–65%. Rains 10–15 days before flower-picking provides excellent flowering and high production, whereas under drought conditions small flowers with small stigmas are expected.

Separation of Stigmas/Styles from Stamens

The separation of stigmas/styles and stamens from the petals is carried out at home within 1 day of collection. First, the flowers are placed in small quantities on a blanket made of goat's wool. Second, with an airstream created by swiftly moving, specially manufactured, leather-bottomed frames in the old days – on which a thin layer of saffron flowers was spread–or by electric ventilators nowadays, petals are separated from stamens and stigmas, which stick to the goat's wool blanket from where they are subsequently collected (Figure 7.6). Next, the red (stigmas and styles) and yellow (stamens) saffron are separated, via one of two methods: (1) by hand, the best but also the most expensive method, used many years ago (2) using a wire screen with 6 × 6 mm holes, which can be either flat or cylindrical. A flat screen yields up to 80% separation, whereas a cylindrical screen yields 90% separation.

Drying

When stamens and stigmas/styles are dried together, the stamens' pollen pollutes and deteriorates the red saffron. It is therefore recommended that they be separated first, before drying. The drying process consists of the following steps: The fresh saffron is placed on 40 × 50 cm trays with a silk-fabric bottom. A thin layer of

Figure 7.4 Hand-picking of saffron flowers at *Krokos*, Kozani, Greece.

Figure 7.5 Baskets for the collection and transport of saffron flowers.

Figure 7.6 Mechanical separator of saffron styles/stigmas and stamens from petals, using an air blower. Kozani, Greece.

saffron (4–5 mm) is spread along this fabric and then these trays are piled on frames with shelves 25–30 cm apart (Figure 7.7). The frames are then placed in a dark room or in a storage room for drying, heated with a firewood stove, and the room temperature controlled. During the first few hours of the drying process, temperature is maintained at 20°C, it is then raised to 30–35°C. The drying process is terminated when the moisture content of the product has been reduced to 10–11%, usually after 12 h. If the red (stigmas and styles) and yellow (stamens) saffron are still together after drying, they can be separated at this stage. At the same time, all foreign substances (soil, hairs, threads, etc.) are removed from the dried saffron product (Figure 7.8). The pure dried saffron is kept in hermetically sealed glass vases or tin cans at 5–10°C.

Figure 7.7 Spread of saffron on tray stacks for drying.

Figure 7.8 Pure dry commercial saffron product (styles and stigmas). Saffron Growers' Cooperative of Kozani, Greece.

Production

To produce 1 kg of fresh red saffron (stigmas and styles), one needs around 80 kg of fresh flowers. However, to produce 1 kg of dried red saffron one needs 120,000–150,000 flowers, or 5 kg of fresh stigmas and styles. The yield of dry red saffron largely depends on weather and soil conditions and the culture treatments the crop has received. In a high-quality plantation, the following annual yields, expressed in dried red product, are expected:

In the first year after planting	3 kg per ha
In the second year after planting	10 kg per ha
In the third and fourth years after planting	15 kg per ha per year
In the fifth and sixth years after planting	10 kg per ha per year

On average, a hectare, within 6 years, produces (a) 60 kg of red saffron (stigmas and styles), (b) 20 kg of yellow saffron (stamens).

TECHNOLOGICAL FEATURES OF GREEK SAFFRON

Main Chemical Ingredients

The main chemical ingredient contained in Greek saffron's drug is protocrocine. After oxidation, this substance produces: (a) two molecules of picrocrocine – $C_{16}H_{26}O_7$ and (b) one molecule of crocine – $C_{44}H_{64}O_{24}$.

Picrocrocine is a glucoside, which upon enzymatic hydrolysis liberates its non-sugary part, which in turn, after oxidation, produces safranal and D-glucosine. These substances are the main constituents of saffron's essential oil, to which saffron owes its characteristic smell. The non-sugary part of crocine, crocetine, is the main dyeing substance to which saffron owes its special red colour. Other substances that exist in saffron are glucomicine, carotene β, p, c, etc.

The final commercial product reaching the market has the following composition (chemical analysis by the Koying method: water; starch; oils; fat; N-substances; non-N-substances; fibres and ash).

Saffron Analysis

The composition of Greek saffron is revealed by several analyses:

Moisture (at 103°C)	8.4–9.6%
Colouring power (440 nm)	120–150
Total ash	5.1%
Essential oil content	1.01–1.12%
Impurities	–
Foreign colours	Absent
Quality difference	Absent

This analysis reflects the excellent quality of Greek saffron.

Characteristics and Uses

Commercial saffron is a natural colouring and aromatic substance derived from fresh stigmas after appropriate drying. Some of saffron's main uses are to improve the colour, smell and taste of many dishes. In small quantities, it stimulates appetite, facilitates digestion and generally strengthens the human organism. Due to its important properties, it is the subject of advanced scientific research to explore its pharmaceutical potential and properties, as these have been reported from the ancient and recent past. Today in the European kitchen, saffron is widely used as a condiment in a variety of food preparations such as rice dishes, pastas, soups (like French bouillabaisse), cakes, saffron bread and numerous sweets. Saffron is also used in the food industry to dye and perfume rice, pastas, candies, dairy products and alcoholic beverages, as well as pharmaceutical products. Other uses of saffron are related to religious ceremonies (India) or to dyeing expensive textiles.

PROFITABILITY OF THE CROP AND AGRICULTURAL INCOME

Saffron cultivation is important for both the growers of Kozani county, in terms of their farm income, and for the Greek agricultural economy – since all annual domestic production is exported.[3]

---

[3] Greek restaurants do not offer saffron dishes (editors' note).

The annual production cost of 1 ha of saffron within the span of a 6-year, economically productive life is $4800 of which $3600 represents human labour. The gross annual income derived from the cultivation of 1 ha of saffron is $5316. The net annual income derived from the cultivation of 1 ha of saffron is therefore $516. This income was calculated on the basis of the actual cultural expenses and the market price of saffron, which was $600 per kg in 1995. A farmer's annual income from the cultivation of 1 ha of saffron is $3701. This income includes the rent for the land ($480 per ha), the net income ($516 per ha) and 75% of the calculated labour cost ($2705 per ha).

A comparison of the annual revenue of the saffron crop in 1995 with the annual agricultural income produced by 1 ha of land covered by other crops in Kozani county leads to the conclusion that saffron is much more profitable. Consequently, saffron could be used as an alternative crop in this region, providing an effective solution to the problem of substituting less profitable crops with more profitable ones to raise the limited agricultural income of the farmers in this county.

The annual saffron growing data for the period 1985–95, including average production, total production and market prices per kg, appear in Table 7.2.

PRODUCT MARKETING AND DISTRIBUTION

Saffron is marketed and distributed by the Saffron Growers' Cooperative of Kozani county. According to the Foundation Law 818/1981 establishing the cooperative, growers are obliged to deliver all of their product to the cooperative every year to secure its joint marketing.

The product is collected from January to late March. After drying and cleaning, it is brought to the cooperative's storage room where, after careful inspection, it is accepted and subsequently stored. Quality is strictly controlled by a panel of specialists, then the product is weighed and packaged by the cooperative in and small 1-g, 2-g, 4-g and 28-g packages or in large-capacity (3 kg) metal cans.

The commercial saffron product is available either as threads or as finely ground powder, which is placed in clean, hermetic packaging consisting of material which excludes the infiltration of foreign substances and loss of the substances it contains.

The package label provides information on: (a) the botanical and commercial name of the product, (b) its net weight and quality category, (c) the country and area

Table 7.2 Data on saffron cultivation in Greece. Mean annual values for the period 1985–95.

Year	1985	1986	1987	1988	1989	1990	1991	1992	1993	1994	1995
Cultivated area in ha	1088.5	1088.5	849.3	849.3	849.3	849.3	770.6	881.6	881.6	877.3	821.3
Mean yield in kg per ha	4.681	5.475	5.556	5.523	5.104	5.625	6.374	4.983	6.922	6.831	8.401
Total annual production in kg	5095.4	5959.5	4718.5	4690.7	4335	4777.7	4911.5	4393.3	6102.7	5993.2	6900
Prices per kg in US $	272	364	480	536	460	448	564	508	472	524	600

of its production, and (d) the recommended expiry, date for its use. The cooperative makes efforts to find customers in foreign markets and to increase saffron consumption in the internal market.

The product is sold internationally and is delivered by an air-transport agency to the destination airport (CIF) from where it is claimed by the recipient following presentation of a Bill of Loading and payment of its value against documents. The transaction is mediated by a Greek bank and a foreign bank designated by the buyer.

Because it has managed to concentrate total production every year, and to promote and sell the product abroad at international prices, (a) the cooperative re-established the trust of serious foreign commercial firms in Greek saffron, (b) the state authorities have begun to show some interest in supporting the crop, (c) the trust of the saffron growers in their cooperative is firmly founded and (d) cultivation has expanded to other areas and villages of Kozani county.

REFERENCES TO GREEK PUBLICATIONS ON SAFFRON

Dodopoulos, S. (1977) *Cultivation of Saffron*. Athens.
Goliaris, A. (Unpublished date).
Kritikos, P. (1960) *Crocus*. Athens.
Papanikolaou, A. (1971) *Saffron*. Thessaloniki.
Tahmatzidis, P. (1980) *Crocus of Kozani*. Kozani.

8. SAFFRON CULTIVATION IN MOROCCO

AHMED AIT-OUBAHOU and MOHAMED EL-OTMANI

*Department of Horticulture, Institut Agronomique et Vétérinaire Hassan II,
B.P. 121, Aït Melloul, 80150 Agadir, Morocco*

ABSTRACT Saffron, a well-known, highly priced dried spice, is obtained from the stigmas and the styles' tops of *Crocus sativus* L. In Morocco, its cultivation is limited to the Taliouine area, located between the cities of Taroudant and Ouarzazate and situated at the junction of the High and Low Atlas mountains. It is an important source of income for many families and is grown on very small plots of land. Currently, saffron acreage covers about 500 ha for a production of about 1000 kg per year, giving an average yield of 2 to 2.5 kg per ha. It is harvested in the autumn months and is marketed in the main urban centres of the country, with a limited amount exported to Europe. This chapter describes the technical aspects of saffron cultivation, harvesting and marketing in Morocco.

INTRODUCTION

Saffron (*Crocus sativus* L., Iridaceae) is a geophyte that propagates solely via annual corms (Mathew 1982). It is a sterile triploid (2n = 24) and is incapable of producing fruit or seeds (Mathew 1977). After drying, its orange-red stigmatic lobes constitute the true saffron spice (Basker and Negbi 1983, 1985; Giaccio 1990) which contains pigments (crocin), an odour (safranal) and a bitter agent (picrocrocin). Its uses are varied and include perfumes, dyes, incense, cosmetics, and medicine (Basker and Negbi 1983). A single flower bears 5 to 7 mg of spice and maximum yields range from 2.5 kg per ha in Kashmir (Bali and Sagal 1987) to 15 kg per ha in Italy (Tammaro and Di Francesco 1978). Saffron is thought to have originated in Greece, Asia Minor and Persia (Skrubis 1990). Today it is cultivated mainly in Spain, Greece, India, Morocco and, to some extent, in several other Asian countries.[1]

BOTANICAL AND BIOLOGICAL CHARACTERISTICS

Crocus sativus L. is named "Zaâfaran" in Morocco, and is a perennial plant having a depressed globule-shaped underground corm, 3 to 5 cm in diameter. The leaves

---

[1] See table 3 in Negbi's article in this volume.

are narrow, six to seven per corm, and grass-like with an elongated blade reaching a length of 30 cm, produced from October until May. Concomitant with leaf production, growth of roots and daughter-corms takes place. The flowers (up to 12 per corm) bloom before leaf emergence. They consist of violet petals which expand at the top. The pistil consists of an ovary from which a pale yellow, slender style arises and divides into a three-lobed stigma which is orange-red and 2.5 to 3 cm long.

The plant goes through a long rest period in summer and active growth in the fall. It also exhibits a short growth period in spring and an even shorter one in winter (Tammaro 1990).

GEOGRAPHICAL, SOIL AND CLIMATIC CHARACTERISTICS OF TALIOUINE

In Morocco, saffron is mainly grown in Taliouine, which is located at the junction of the Low Atlas in the South and the High Atlas mountains in the North, at a latitude of 30° 36'N, a longitude of 8°25'W, and an altitude of about 1000 m (see map, Figure 8.1). However, saffron-growing plots are located at an altitude of about 1200 to 1400 m in a warm microclimate which is nevertheless cooler than in Taliouine itself, thereby enabling higher yields and better quality saffron. Winter is relatively cold with snowy days. Although the snow cover is usually thin and short-lasting, it can cause significant damage to the leaves, the photosynthetically active organs. This, in turn reduces the filling of the daughter corms and, consequently the yield of the following year. Summer temperatures can reach 25–30°C and air is relatively dry year-round. Rainfall in the area ranges from 100 to 200 mm per

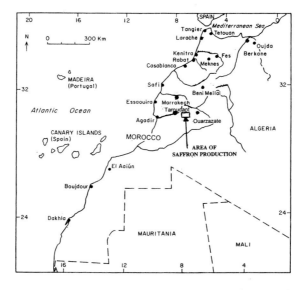

Figure 8.1 Map of Morroco indicating Taliouine area where saffron is cultivated.

annum. Dominant winds blow in a N–NW direction and frost can occur from January to March. Soils in the saffron-growing areas are either sandy loam or calcareous clay with a fairly loose texture. The latter type is dominant in the counties of Taliouine, Zagmouzen and Agadir Melloul.

PROPAGATING MATERIAL

Corm formation and filling occur during the period of vegetative growth (October–March). In March, leaves are cut back and corms undergo a natural dormancy period.

During August and September, the corms, which are onion-shaped and covered with fibrous tunics, are dug up. The daughter corms are separated and subjected to a selection process based on the elimination of rotten, bruised or damaged corms. From the remaining corms, the external 2 to 3 tunics are removed, leaving only the interior one. Only corms with a diameter greater than 2.5 cm are used in propagation. The rest are used as animal feed. Small-size corms can result from crowding in the soil, leaf loss or damage before full corm development and maturation, or severe drought during corm formation. Corms can be stored for several weeks in a cool, dry environment, but better sprouting is obtained if they are used shortly after having been dug up.

FREQUENCY OF RENEWAL OF A SAFFRON PRODUCING PLOT

After flowering the parent corm gives rise to two to three new corms which develop to replace the parent. On a given saffron plot, this process continues for several years: every year a corm develops new daughter corms, which can end up crowding each other until insufficient space is left for the corms to develop to a sufficiently large size to yield a good harvest. Moreover, every year daughter corms usually ascend about 1 to 2 cm higher in the soil than those of the previous year, until they end up reaching the soil surface. Again, yields decrease significantly and at that stage, the corms have to be uprooted and moved to a new plot. Under the conditions of Taliouine, a given saffron planting can keep producing for 5 to 12 years, on average, depending mainly on the planting density. Higher density plantings give greater yields but need to be replaced within a few years. Between plantings, the field is cultivated for about 3 to 5 years with other crops, mainly cereals, vegetables and legumes.

LAND PREPARATION AND PLANTING

Because of the mountainous terrain, saffron is planted on terraces made on the hill slopes (Figure 8.2). Planting plots are seldom larger than 100 m^2 and may already be planted with almonds, olives or other fruit trees (Figure 8.3). Before planting, the soil is thoroughly cleared of undesirable material, ploughed to a depth of about 30 cm and then left to rest for several weeks. Ploughing is performed mainly using manual labour and animals.

Figure 8.2 Terraces where saffron is grown.

Figure 8.3 Saffron grown under fruit trees.

Farm manure is used at 10 to 20 tons per ha, depending on its availability, and thoroughly mixed with the soil. Before planting, the land is levelled to facilitate irrigation. The soil is prepared during August and September. Manure is also used in the third or fourth year of planting. No other fertilizers are used in the area for saffron production.

Because the area has a relatively dry climate, it is relatively free of diseases which attack saffron. The corms are therefore not subjected to any treatment before planting. However, in a few plantations in the Taliouine area, the corms are infected by the nematode *Ditylenchus dipsaci* (Radouni 1985). Nevertheless, the negative effect of this nematode on corm formation and yield is limited.

Before planting, the land is divided into small, 2 m × 2 m plots, each of which is irrigated separately. Planting is done manually, in rows which are 20 cm apart. Bunches of two to three corms each are planted 10 to 15 cm apart within rows. Planting depth is about 15 cm and about 3 tons of saffron corms are used per hectare. Planting takes place in late August to September. Corms are stored, if necessary, for a short period of time in plastic bags.

After planting, the crop is irrigated and the soil is superficially worked once it is relatively dry to allow for good aeration and flower emergence.

OTHER CULTURAL PRACTICES

Irrigation

Terraces are irrigated by flooding the plots with water, which is either drawn from natural mountain springs using canals and basins for storage and distribution, or pumped from 18 to 50 m deep wells. Irrigation, 350–500 m^3 of water per ha, is performed once a week from September to November and every other week from December to March. No irrigation is done during the months of April–August which correspond to the period of deep corm dormancy.

Weed Control

Weeds are a problem in saffron plantations during the plants' growth period (October–April). They are controlled by hand only, as herbicides are unknown to the growers. The weeds are used as feed for animals.

Flower Harvesting

Flowering takes place in the fall (mid-October to end of November), with the greatest concentration of flowering occurring during the first 10 days of November. The flowers are ephemeral and are normally harvested the same day they emerge. Flower picking is laborious and performed mainly by family members. When outside labour is used, the pay is one-tenth of the amount of saffron harvested by that labour. The task includes picking the flowers and separating the stigmas from the petals and stamens.

Flower picking takes place early in the morning, while the flower is still closed, before the perianth segment opens. Flowers are picked at the base of the segments, and put into baskets in thin layers to avoid excess pressure and deformation of the flower organs, particularly of the stigmas. Picking goes on for the first two to three morning hours. Immediately after harvest, the flowers are brought indoors for separation. During the process, the stigmas plus the uppermost 2 mm of the style are separated from the rest of the flower organs. If the style portion is longer than 2 mm, saffron is considered to be of inferior quality.

During the flower-picking period, because the flowers have to be collected early in the morning before they open, as many workers as possible are needed to finish this operation as quickly as possible. It is generally believed by the farmers that saffron from flowers that have been open for long periods in the field is of inferior quality. This was demonstrated by Zanzucchi (cited in Tammaro 1990: 53, 73) who found reduced flavouring and colouring power of saffron from open vs. closed flowers.

Drying of Saffron

The fresh red stigmas are dried immediately after harvest. Saffron is handled very gently and carefully to avoid stigma breakage and to ensure optimum conditions for the development of a prime quality product. The stigmas are placed on a cloth in thin layers and dried in the sun for a 2 h period or in the shade after 7 to 10 days. Drying is complete before the stigmas break or crumble. Air-dried saffron retains its purplish red colour, its fragrance and its aroma, and commands a high price in the market place.

Yield

Yield is relatively low in the first year and increases to maximum in the third to fifth years after planting. After this, flower production may decrease. Saffron can vary from 2 to 6 kg per ha, based on planting density, plantation age and climatic conditions during the previous year. The average yield in the country varies from 2 to 2.5 kg per ha, which is very low in comparison to modern saffron plantations in Spain or Italy; lack of rain and irrigation during corm formation and plant growth significantly reduces the yield. One kg of intact flowers yields 72 g of fresh saffron (stigmas), which in turn yields 12 g of dry saffron. The final product retains about 5–20% humidity.

Storage of the Dried Product

In Morocco, saffron is stored as whole dried stigmas and seldom as a powder. Dried saffron is highly hygroscopic; it is kept in well-sealed clay jars or coloured glass containers, or in tightly closed tin cans, and stored in a dark, dry and cool place. It can also be stored in tightly closed, dark plastic bags in a dry environment for many years.

Uses, Marketing and Economics

At the present time, saffron is sold locally to a cooperative or to buyers that serve as intermediaries between the grower and the wholesaler. The cooperative and the buyers sell the product to wholesalers usually located in the main cities of the country such as Casablanca, Rabat, Fès and Marrakech. From there, saffron is either sold in local markets or transferred to other cities. A very small portion is exported, mainly to France and Spain, which in turn re-export some of the quantity to other European countries.

Although the quantity of saffron produced in Taliouine is small, it is a biological product that deserves the highest attention because of its excellent organoleptic qualities. Saffron is highly valued for these qualities and considered a source of income for over 2000 families.

In Morocco, use of saffron is mostly limited to food preparation or hot drinks, especially green tea. It adds colour and flavour to all kinds of traditional dishes, to mention only a few, such as couscous, tajine, pastilla, harira and pastries.

RESEARCH AND DEVELOPMENT PROGRAMMES

Current yields (2 to 2.5 kg per ha) are very low compared to those obtained in Italy (10–16 kg per ha) or Spain (10–12 kg per ha). Efforts are being devoted to better understanding the factors limiting yield and to improving the existing or introducing new cultural practices (use of fertilizers, selection of plant material, irrigation, etc.). Moreover, to improve the crop's benefit to the grower, organization of the marketing sector is under way, with the creation of a cooperative (with approximately 62% of the total surface area of saffron) being only the beginning. This reorganizing, coupled with a good advertising programme locally and overseas, is expected to ensure better marketing and sales. This would bring a better return to the farmer and would contribute to increasing interest in saffron cultivation and agriculture in general among young people.

Experiments to adapt the cultivation of saffron to other areas where its culture might be possible, such as Zagora, located south of Ouarzazate are being carried out. Other experiments are being carried out in other locations such as Errachidia. Results have not always been conclusive and research continues.

REFERENCES

Bali, A.S. and Sagwal, S.S. (1987) Saffron – a cash crop of Kashmir. *Agricultural situation in India*, March 1987, 965–968.

Basker, D. and Negbi, M. (1983) The uses of saffron. *Economic Botany*, **37**, 228–236.

Basker, D. and Negbi, M. (1985) Crocetin equivalent of saffron extracts. Comparison of three extraction methods. *J. Assoc. Public. Anal*, **23**, 65–69.

Giaccio, M. (1990) Components and features of saffron, in Tammaro, F. and Marra, L. (eds.) (1990) *Lo Zafferano: Proceedings of the International Conference on Saffron (Crocus sativus* L.) L'*Aquila* (Italy) 27–29 October 1989. Universitá Degli Studi L'Aquila e Accademia Italiana della Cucina, L'Aquila, pp. 135–148.

Mathew, B. (1977) *Crocus sativus* and its allies (Iridaceae). *Plant Systematics and Evolution*, **128**, 89–103.

Mathew, B. (1982) *The crocuses: a revision of the genus Crocus*. B.T. Batsford, London.

Radouni, J. (1985) Monographie du Saffran dans la région de Taliouine. Mémoire d'ingénieur d'application, IAV Hassan II, Agadir, Maroc.

Skrubis, B. (1990) The cultivation in Greece of *Crocus sativus* L., in Tammaro, F. and Marra, L. (eds.) (1990) *Lo Zafferano: Proceedings of the International Conference on Saffron (Crocus sativus* L.) L'*Aquila* (Italy) 27–29 October 1989. Universitá Degli Studi L'Aquila e Accademia Italiana della Cucina, L'Aquila, pp. 171–172.

Tammaro, F. (1990) *Crocus sativus* L. cv. Piano di Navelli (L'Aquila saffron): environment, cultivation, morphometric characteristics, active principles, uses, in Tammaro, F. and Marra, L. (eds.) (1990) *Lo Zafferano: Proceedings of the International Conference on Saffron (Crocus sativus* L.) L'*Aquila* (Italy) 27–29 October 1989. Universitá Degli Studi L'Aquila e Accademia Italiana della Cucina, L'Aquila, pp. 47–57.

Tammaro, F. and Di Francesco, L. (1978) *Lo zafferano dell'Aquila*. Istituto di Tecnica e Propaganda Agraria, Roma, 20 pages.

9. SAFFRON TECHNOLOGY

DOV BASKER

*Department of Food Science,
Agricultural Research Organization,
The Volcani Centre, Bet Dagan, Israel*

ABSTRACT Before marketing, saffron stigmas must be picked, separated and dried. The latter step, known in peasant cultures, has not yet been defined nor optimized scientifically. Dyeing with saffron is discussed, as are its milling and packaging for use as a food spice.

Etymological note: The oldest known document in which the source-word of "saffron", za'afran, appears is the Jerusalem Targum (ca 250 CE) in Judeo-Aramaic, a dialect of the dominant west-Semitic language in the Near East at that time. No technological details were given.

INTRODUCTION

All technology is "easy" – provided that somebody else is doing it. We only begin to understand the problems overcome by the craftsman, be he tinker or tailor, when we try to duplicate his art without his guidance. How difficult can it be to dry saffron? Then why is the quality, as reflected in the market price, so variable among producers (Sampathu et al. 1984; Ward 1988; Vinning 1989)? The best Spanish saffron is well known in world markets and indeed dominates them; only *its* price is quoted in the general press, and this depends on supply alone (Basker 1993): in the range of 17 to 118 tonnes *per annum* (mean = 39, median = 26) exported from Spain from 1975 to 1986, the last data available,

$$\text{mean wholesale price, (1991) US \$ per kg} = 10^{4.7} * (\text{tonnes})^{-1.20} \qquad [1]$$
$$(N = 11, \ R = -0.95, \ d.f. = 9, \ p<<0.001)$$

The retail price, in perhaps 1- or 2-g packages, is several-fold higher. Other growers cater to their local markets but export only relatively small amounts.

QUALITY

"Quality" is made up of many factors, beginning with guaranteed authenticity and secondly the assurance of non-adulteration. Both these factors can be checked by laboratory examination, essential when confronted with a supplier of unknown history, but they are not sufficient. The next major parameter (see the chapter on saffron chemistry, this volume) is the colour intensity, followed by odour and taste. Retail consumers can make their own subjective assessments of the sensory parameters: place 5 to 10 mg of dried saffron in a white (internally) 250-ml cup, add boiling water to fill, and note the colour extracted, the odour and (after some cooling) the taste; for comparison, the three dried stigmas from a single flower's triad weigh 5 mg (Basker 1993), and the pharmaceutical dose "for those at death's door" would be 500 mg (Gerard 1633; Culpeper 1652), while 1.5 g (1500 mg) may be fatal (Fasal and Wachner 1933). To make "saffron tea" (Quimm 1976), add a (1 g) teabag briefly to the hot saffron extract before cooling and add sugar, sweetener or lemon to taste, but not milk.

PICKING

Picking crocus flowers requires intensive stoop-labour (Tammaro 1990): the flowers are only a few centimetres above-ground and are frequently, but not always (Negbi 1990), surrounded by long thin leaves which must not be damaged – or else daughter-corms will not form to replace the current generation. Flowering takes place in autumn, lasts only two to three weeks (Basker 1993) and picking is required almost daily (*ibid.*; Anon. 1982; Tammaro, this volume): the flowers wilt rapidly and once this has happened they cannot readily be separated into their constituent parts. Reports that whole flowers are "dried in the sun as such" and the stigmas later "picked by hand" should not be given much credence if the product is expected to be saffron in more than just name.

SEPARATION

Cutting the style with its three attached stigmas is followed by physical separation from the petals and anthers. These stages are also traditionally labour-intensive but can be performed while seated in comfortable surroundings (Greenberg and Lambert Ortiz 1983, photograph on pages 76–77; Ward 1988). One common method of doing both simultaneously is to pluck the stigmas from the flower (Sampathu *et al.* 1984; Tammaro 1990), or with a fingernail. Mechanical cutting is possible theoretically but difficult in practice, to be followed by fan-separation (Galigani 1987; Kapur 1988; Skrubis 1990). A third reported separation method consists of water-flotation (Watt 1908; Nauriyal *et al.* 1977), almost guaranteed to result in a low-quality product because of leaching of water-soluble constituents, including the important colour parameter.

DRYING

Three market forms of saffron are known: "hay", "cake" and powder. The drying process employed is always the most critical with regard to quality (Zarghami 1970;

Pruthi 1980). The relationship between temperature and the time required to dry hay saffron – stigmas in the loose state (*Oxford English Dictionary* 1971) – to completion has been studied (Basker 1993), but use of the resultant product to estimate optimal-drying sensory conditions failed (*ibid.*) because, with hindsight, all volatiles had been lost. Proximate analysis of commercial saffron also shows that drying to zero moisture is not an appropriate model (see Table 1 in the chapter on saffron chemistry, this volume). Zarghami (1970) recommends "proper roasting" at unspecified low temperatures. Ameloti and Mannino (1977) write positively of a "fermentation process" [as opposed to a negative connotation for fermentation during storage (Tammaro 1990)] but do not elucidate. Charcoal fires (Tyler *et al.* 1976; Sokolov 1989; Skrubis 1990; Tammaro 1990) are used for "artificial heating" (Zarghami 1970; Sampathu *et al.* 1984) with few practical details save a remark that "too much heat" destroys the aroma (Ward 1988).

Solar drying, in sun or in shade, has been used as an alternative to "artificial heating" (Watt 1908; Moldenke and Moldenke 1952; Morton and Zallinger 1976; Nauriyal *et al.* 1977; Kapur 1988) even though it is almost guaranteed to result in a photochemical decrease in colour intensity (see the chapter on saffron chemistry, this volume). Drying by solar exposure may be "natural", but as the resultant product shows, it is also crude; the constraint does not apply if sunlight is used only as the heat source, without exposure.

The third drying method, apparently no longer in use, is for the production of cake saffron (Rosengarten 1969; *Oxford English Dictionary* 1971). For this method, a layer of stigmas approximately 6 cm thick was "kiln-dried" under the pressure of a board (Howard 1678; Douglass 1729; *Encyclopaedia Britannica* 1905; Grieve 1959), first for 2 h at one unstated temperature and for a further 24 h at a lower one, also unstated. Our own trials were unsuccessful, possibly because insufficient raw material was available to build up a layer of adequate thickness. The statement that honey and safflower are added to the saffron cake (Wren 1980) must be treated warily; use of honey may be justified technologically as a binder (doubtful), if permitted by local legislation and properly declared on the label; but the addition of safflower could constitute a *prima facie* case of adulteration. The final product has been described as a "compressed matted mass" (Rosengarten 1969).

DYE

Saffron has been used as a yellow dye for wool since ancient times (Basker and Negbi 1983), although Liddell and Scott (1897) feel that this was not yet so in the Homeric era (8th century BCE). A statement that the colour's water-solubility renders it unsuitable for dyeing (Stuart 1979) indicates that the author was unaware that the material to be dyed must be mordanted, *i.e.* soaked in a warm solution of alum [$K_2Al_2(SO_4)_2$] and cream of tartar [KH tartrate] and then dried (Schetky 1968) before beginning the actual dyeing procedure. The result is a long-lasting yellow. It was once used in Turkish carpets, retaining the colour after decades of constant use (A. Kempinsky 1983, pers. comm.); for economic reasons, saffron is no longer used for this purpose (*ibid.*). Saffron was also one of the first products used to stain biological tissues for microscopy (Lewis 1942; Gurr 1956; see the section on Histochemistry in Negbi, this volume).

MILLING

Saffron is milled into a powder by equipment which shears the brittle dry stigmas (Tammaro 1990). Crushing, as in a ball-mill or with a mortar and pestle, does not produce the desired result with any efficiency. Some consumers are wary of purchasing powdered saffron, fearing the ease with which it may conceal an adulterant from the naked eye, and so prefer hay saffron; on the other hand, colour extraction from the powder may be more complete (International Standards Organization 1980), and milling may also release odoriferous material entrapped in the tissue and/or result in thermal dissociation of picrocrocin to release safranal. Hay saffron may be crushed in the presence of salt or sugar (Roden 1975), which for marketing would require the permission of local legislation and label-declaration; with ingredients whose cost differs by several orders of magnitude, permission for such admixture might presuppose inordinate one-sided official benevolence.

PACKAGING

The small amounts of saffron in individual retail packages, usually less than 2 g, result in problems for non-specialized packers whose equipment is insufficiently precise. Instances of serious short- and over-weight have been noted – the former liable to result in prosecution and the latter in financial loss – and in one case an automatic weigher + printer marked packages as containing "0.000 kg"! Specialized packers have the means of maintaining net weight to within less than 0.1 g tolerance, no small engineering feat and requiring a throughput sufficient to justify the investment in equipment; these packers do not use simple multi-purpose transparent bags, capsules or containers for their product, but custom-designed and hermetically sealed laminates to protect against air (oxygen), light, moisture, contamination and quality deterioration, as well as to retain volatiles such as safranal; each such envelope may then be inserted into a much large container for retail presentation as part of a range of similar packs for other spices, more bulky and less expensive per unit weight than saffron.

Until a wholesale of package – say, 10 kg – of saffron is prepackaged for retail sale it can be stored under lock and key with relative ease, provided that due precaution is taken to protect it against quality deterioration (Tammaro 1990), including cool temperatures, preclusion of air and controlled humidity (Mannino and Amelloti 1977; Alonso *et al.* 1990). Sulphur dioxide (SO_2), which has a wide range of industrial applications, must be scrupulously distanced, even in tiny amounts, from saffron (Rayner 1991) to prevent bleaching. Packaging for retail sale requires time, labour, equipment, relatively large quantities of suitable laminate, containers, labels and cartons – and appropriate finance and a market network: the same wholesale package may result in, say, 200 cartons with 5,000 packs of 2 g net each, presenting a new set of warehousing problems to be overcome.

MARKETING

Expansion of the market for saffron spice in home or restaurant food preparation does not appear to be linked to any increase in general affluence within cultures without preknowledge of its use (Olney 1977). In 1986, for example, the US, with a vastly greater population, imported only about as much saffron as did Sweden (Commonwealth Secretariat, London 1988, pers. comm.). Economic reality limits its use on humbler tables, unless home-grown (Claire 1961). Newer uses in processed foods presuppose that sensory and subjective quality will be emphasized over purely economic considerations by consumers as well as by advertisers. Saffron has been proposed for mayonnaise (Carrier 1976) and soft drinks (Timberlake and Henry 1986), and is known to be used in the limited opulent market for halwa (Dagher 1991). Methods for confirming its presence in such foods have been investigated (Montag 1962; Corradi et al. 1981). However, little purpose would be served by adding saffron to oil-based foods such as butter (Basker and Negbi 1983) as α-crocin is not oil-soluble; in flour confectionery there is some risk of a green hue forming by reaction with the enzyme β-glucosidase over a 6- to 8-h period in the presence of proteins, but the enzyme is inactivated during baking (Rayner 1991). It has also been suggested that carotenoid products might find uses in cosmetics (Bauernfeind et al. 1971), where marketing costs often swamp ingredient costs.

A recipe for "Saffron Broth" was included in the first cookbook to be printed with movable type (Platina 1475): 30 egg yolks and unstated quantities of unripe grape juice, veal or capon juice, saffron and cinnamon were to be mixed, strained and cooked, stirring until thickening occurred, and sprinkled with (undefined) spices for ten servings.

REFERENCES

Alonso, G.L., Varón, R., Gómez, R., Navarro, F. and Salinas, M.R. (1990) Auto-oxidation in saffron at 40°C and 75% relative humidity. *Journal of Food Science*, **55**, 595–596.

Amelotti, G. and Manning, S. (1977) [Analytical evaluation of the commercial quality of saffron.] *Revista della Società Italiana di Scienza dell' Alimentazione*, **6**, 17–20 (in Italian).

Anonymous (1982) *Spices, A survey of World Trade*. International Trade Centre, UNCTAD/ GATT, Geneva, Vol. 1, pp. 71–72.

Basker, D. (1993) Saffron, the costliest spice: drying and quality, supply and price. *Acta Horticulturae*, **344**, 86–97.

Basker, D. (1999) Saffron chemistry (this volume).

Basker, D. and Negbi, M. (1983) The uses of saffron. *Economic Botany*, **37**, 228–236.

Bauernfeind, J.C., Brubacher, G.B., Klaui, H.M. and Marusich, W.L. (1971) Uses of carotenoids. In O. Isler (ed.), *Carotenoids*, Birkhauser Verlag, Basel, pp. 743–770.

Blacow, N.W. (Ed.) (1972) *Martindale. The Extra Pharmacopoeia* (26th ed). The Pharmaceutical Press, London, pp. 726.

Carrier, R. (1976) *The Robert Carrier Cookery Course*. Sphere Books Ltd., London, Vol. 1, pp. 60.

Clair, C. (1961) *Of Herbs and Spices*. Aberland-Schuman, London, pp. 15.

Corradi, C., Micheli, G. and Sprocate, G. (1981) [Determination of saffron in compound food products by identification of colour, bitter taste and odour.] *Industrie Alimentari* (in Italian), **20**, 624, 627– 629.

Culpeper, N. (1652) Quoted by Silberrad and Lyall (1909).

Dagher, S.M. (Ed.) (1991) *Traditional Foods in the Near East*. FAO Food and Nutrition Paper 50, Food and Agriculture Organization, Rome, pp. 151.

Douglass, J. (1729) An account of the culture and management of saffron in England. *Philosophical Transactions*, **35**, 566–574.

Encyclopaedia Britannica (1905) The Werner Company, Akron, OH, Vol 21, pp. 153–154.

Fasal, P. and Wachner, G. (1933) *Wein. klin. Wschr.* **45**, 745. Quoted by Blacow (1972).

Galigani, P.F. (1987) La meccanizzazione delle colturie di salvia, lavanda, zafferano e genziana. In A. Bezzi (1987) *Atti Convegno sulla Coltivazione delle Piante Officinalli*, Trento, 1986, Ministero dell'Agricoltura e delle Foreste, pp. 221–235 (Italian, English summary).

Gerard, J. (1633) The Herball or Generall Historie of Plantes, reprinted 1975. Dover Publications Inc., New York, p. 154.

Greenberg, S. and Lambert Ortiz, E. (1983) *The spice of life*. Michael Joseph/Rainbird, London.

Grieve, M. (1959) *A Modern Herbal*. Hafner Publishing Co., New York, Vol. 2, pp. 698.

Gurr, E. (1956) *A Practical Manual of Medical and Biological Staining Techniques*. Leonard Hill [Books} Ltd., London, pp. 194.

Howard, C. (1678) A relation of culture, or planting and ordering of saffron. *Philosophical Transactions* **12**, 945–949.

International Standards Organization (1990) *Saffron – Specification*. International Standard 3632, Geneva.

Jerusalem Targum, *VaYikra*, ch. 15, v. 19., M. Ginsburger (transcriber) (1903). S. Carvary & Co. Berlin, p. 199.

Kapur, B.M. (1988) *Annual Report* 1987–88, Regional Research Laboratory, Council of Scientific Industrial Research, Jammu, pp. 56.

Lewis, F.T. (1942) The introduction of biological stains: employment of saffron by Vieussens and Leeuwenhoek. *Anat. Rec.*, **83**, 229–253.

Liddell, H.G. and Scott, R. (1897) *A Greek–English Lexicon* (8th ed). American Book Company, pp. 847.

Mannino, S. and Amelotti, G. (1977) [Determination of the optimum humidity for storage of saffron.] *Revista della Societa Italiana di Scienza dell' Alimentazione* **6**, 95–98 (in Italian).

Moldenke, H.N. and Moldenke, A.L. (1952) *Plants of the Bible*. Chronica Botanica Company, Waltham, MA, pp. 87.

Montag, A. (1962) *Z. Lebens. Unters. Forsch.* **116**, 413 ff. Quoted by Banerfeind *et al.* (1971).

Morton, J.F. and Zallinger, J.D. (1976) *Herbs and Spices*, Golden Press, New York, pp. 98.

Nauriyal, J.P., Gupta, R. and George, C.K. (1977) Saffron in India. *Arecanut and Spice Bulletin*, **8** (3), 59–72.

Negbi, M. (1990) Physiological research on the saffron crocus (*Crocus sativus*). In F. Tammaro and L. Marra (eds.), *Lo Zafferano: Proceedings of the International Conference on Saffron (Crocus sativus L.) L'Aquila* (Italy) *27–29 October 1989*. Universitá Degli Studi L'Aquila e Accademia Italiana della Cucina, L'Aquila, pp. 183–207.

Olney, R. (1977) *Simple French Food*. Atheneum, New York.

Oxford English Dictionary, Compact ed (1971). Oxford University Press, Glasgow.

Platina (Bartolomeo de Sacchi di Piadena) (1475) *De Honesta Voluptate* [On Honest Indulgence and Good Health], Andrews, E.B. (trans.), Vol. V. Mallinckrodt Collection of Food Classics, Lib. VI, Cap. (in Latin and English).

Pruthi, J.S. (1980) *Spices and Condiments: Chemistry, Microbiology, Technology.* Academic Press, New York, pp. 182.

Quimm, P. (1976) *The Signet Book of Coffee and Tea.* New American Library, New York, pp. 138.

Rayner, P.B. (1991) Colours. In J. Smith (ed.), *Food Additive User's Handbook.* Blackie, Glasgow, pp. 106, 107.

Roden, C. (1975) *A Book of Middle Eastern Food.* Penguin Books, Hamondsworth, pp. 38.

Rosengarten, F. (1969) *The Book of Spices.* Livingston Publishing Co., Wynnewood, PA, pp. 390.

Sampathu, S.R., Shivashankar, S. and Lewis, Y.S. (1984) Saffron (*Crocus sativus* Linn.) – cultivation, processing, chemistry and standardization. *CRC Critical Review in Food Science and Nutrition*, **20**, 123–157.

Schetky, E.J. (Ed.) (1968) *Dye Plants and Dyeing – A Handbook*, Brooklyn Botanical Gardens, Brooklyn, NY, pp. 19.

Silberrad, U. and Lyall, S. (1909) *Dutch Bulbs and Gardens.* Adam & Charles Black, London, p. 38.

Skrubis, B. (1990) The cultivation in Greece of *Crocus sativus* L. In F. Tammaro and L. Marra (eds.), *Lo Zafferano: Proceedings of the International Conference on Saffron (Crocus sativus L.) L' Aquila (Italy) 27–29 October 1989.* Universita Degli Studi L'Aquila e Accademia Italiana della Cucina, L'Aquila, pp. 171–182.

Sokolov, R. (1989) The fruits of Spanish labor. *Natural History*, March 1989, pp. 82–85.

Stuart, M. (Ed.) (1979) *VNR Color Dictionary of Herbs and Herbalism.* Van Nostrand Reingold Co., New York, p. 52.

Tammaro, F. (1990) *Crocus sativus* L. cv. Piano di Navelli (L' Aquila saffron): environment, cultivation, morphometric characteristics, active principles, uses. In F. Tammaro and L. Marra (eds.), *Lo Zafferano: Proceedings of the International Conference on Saffron (Crocus sativus L.) L' Aquila (Italy) 27–29 October 1989.* Universita Degli Studi L'Aquila e Accademia Italiana della Cucina, L'Aquila, pp. 47–97.

Timberlake, C.F. and Henry, B.S. (1986) Plant pigments as natural food colours. *Endeavour*, 10, 31–36.

Tyler, V.E., Brady, L.R. and Robbers, J.E. (1976) *Pharmacognosy* (7th ed.). Lea and Febiger, Philadelphia, PA, p. 102.

Vinning, G. (1989) *Growth, Production and Distribution of Spices.* Australian Centre for International Agricultural Research, Working Paper No. 27.

Ward, C.R. (1988) Flowers are a mine for a spice more precious than gold. *Smithsonian*, 19 (5), 104–111.

Watt, G. (1908) *The Commercial Products of India.* John Murray, London, p. 430.

Wren, R.C. (1980) *Potter's New Cyclopaedia of Botanical Drugs and Preparations.* Health Science Press, Hengiscote, Bradford, p. 265.

Zarghami, N.S. (1970) *The Volatile Constituents of Saffron (Crocus sativus* L.). Ph.D. Thesis, Univ. Calif. Davis, p. 83.

10. SAFFRON IN BIOLOGICAL AND MEDICAL RESEARCH

FIKRAT I. ABDULLAEV[1] and GERALD D. FRENKEL[2]

[1]*Laboratorio de Oncologia Experimental, Instituto Nacional de Pediatria, Av. Insurgentes Sur 3700-C 04530 Mexico, D.F. Mexico*
[2]*Department of Biological Sciences, Rutgers University, Newark, New Jersey, USA*

ABSTRACT There is a long history of the use of saffron in the traditional medicines of many cultures. As a result of a variety of recent scientific investigations, there is now convincing evidence for the biological activity of saffron and its constituents. One of the activities of saffron which has the greatest potential medical applicability is its ability to inhibit carcinogenesis. A number of recent studies have shown that saffron extract possesses antitumor activity against transplanted tumors and anticarcinogenic activity against chemically induced carcinogenesis *in vivo*, and cytotoxic effects on tumor-derived cells *in vitro*. These findings have raised the possibility that natural saffron and/or some of its constituents might be used as alternative antitumor or anticarcinogenic agents, either alone or in combination with synthetic substances having anticancer activity. The recent scientific findings on the biological activities of saffron, together with the body of anecdotal evidence for its therapeutic activity against a number of diseases, provide strong indications that saffron and/or its components may be useful agents in modern medicine.

INTRODUCTION

In the past few years, there has been increasing interest in the biological effects of saffron and its potential medical applications (see Table 10.1). The scientific literature on this aspect of saffron was the subject of a review which appeared several years ago (Abdullaev 1993). Accordingly, this chapter will focus on recent developments in the field, within the context of the body of previous knowledge.

To a certain extent, the recent interest in saffron is part of a generally increasing awareness of the great medical potential of natural products (such as spices) with low toxicity. In addition, however, the long history of traditional medical applications of saffron suggests that the scientific investigation of saffron and its constituents will prove to be particularly fruitful. Accordingly, before discussing the current scientific literature on the biological effects of saffron, we will briefly review this medical history. It is important to bear in mind that the basis for these medical applications is almost entirely anecdotal; nevertheless, as a whole, these reports can

Table 10.1 Scientific institutions in which research on the biological properties of saffron has been carried out during the past five years.

Institution	Location	References
Amala Cancer Research Center	Kerala, India	Nair et al.
Central Food Technology Institute	Mysore, India	Sujata et al.
University of Minnesota Medical School	Minneapolis, MN, USA	Nair et al.
Agricultural University of Athens	Athens, Greece	Tarantilis et al.
Université de Reims Champagne-Ardenne	Reims, France	Tarantilis et al.
Rutgers University	Newark, NJ, USA	Abdullaev et al.
Azerbaijan Academy of Sciences	Baku, Azerbaijan	Abdullaev et al.
Shanghai Chongming City.	Shanghai, China	Ni et al.
Research Institute, QP Corp.	Fuchu, Japan	Isa et al.
Kyushu University	Fukuoka, Japan	Morimoto et al.
University of Murcia	Murcia, Spain	Castellar et al.
University of Tokyo	Tokyo, Japan	Sugiura et al., Zhang et al.
National Institute of Hygienic Sciences	Ibaraki, Japan	Morimoto et al.
University of Castilla-La Mancha	Albacete, Spain	Escribano et al.
University of Queretaro	Queretaro, Mexico	Abdullaev and Gonzalez de Mejia

reasonably be taken as an indication of the appropriateness of a significant scientific investigation into the potential medical applications of saffron.

Historically, saffron has been employed in many medicinal remedies against numerous conditions. In Chinese traditional medicine, saffron has been widely used for its anodyne, tranquilizing and emetic properties. It has also been used in the treatment of menstrual disturbances, thrombus diseases and some other diseases related to high blood viscosity. It has found applications in nervous disorders: to allay fears, cure trances and in the treatment of some disorders of the central nervous system (Suzhou New Medical College 1977, Zhou et al. 1987, Liakopoulou-Kyriakides and Skubas 1990). Its medical value was recorded in Yi-Lin-Ji-Yao, a traditional Chinese medical book composed during the Ming Dynasty (16th century); notable among the effects described was the promotion of blood circulation to remove blood stasis. The book Yinshanzhengyao ("The Importance of Diet") (circa 1550) contains 136 recipes which include saffron for treating a variety of conditions. Saffron also appears in several traditional Chinese pharmaceutical compendia (Ni 1992). It has been used in traditional Indian and Azerbaijani medicine to treat various diseases including cancer, heart disease, eye disease, blood disease and muscle paralysis (Kasumov 1970, Pfander and Witwer 1975, Nadkarni 1976, Damirov et al. 1988).

A few scientific reports of medico-pharmacological significance have also appeared concerning saffron and its components. Miwa (1954) reported an inhibitory effect on the increase of bilirubin in the blood, and Gainer and Jones (1975) reported a decrease in serum cholesterol and triglyceride levels induced by crocin and crocetin. More recently, saffron extracts have been reported to contain both a platelet-aggregation inducer and inhibitor (Liakopoulou-Kyriakides and Skubas 1990).

BIOLOGICAL EFFECTS OF SAFFRON AND ITS COMPONENTS

Studies *In Vivo*

There have been several recent investigations focusing on the effects of saffron and its components on the nervous system which have led to the discovery of an apparent interaction with ethanol. In one study, Zhang *et al.* (1994) examined the acute effects of saffron extract on passive-avoidance performance in normal and in learning- and memory-impaired mice. A single oral administration of extract had no effect on memory registration, consolidation or retrieval in normal mice. However, the extract did reduce the ethanol-induced impairment of memory registration in both step-through and step-down tests and the ethanol-induced impairment of memory retrieval in step-down tests. The extract also decreased the motor activity and prolonged the sleeping time induced by hexobarbital. The authors suggested that saffron ameliorates the impairment effects of ethanol on learning and memory processes, and possesses a sedative effect. They suggested four possible mechanisms for this effect: 1) saffron facilitates the detoxification of alcohol by decreasing its absorption from the gastrointestinal tract; 2) saffron accelerates the elimination of alcohol from the brain by promoting its metabolism in the liver; 3) saffron accelerates the elimination of alcohol from the brain by promoting blood circulation; and 4) saffron antagonizes the pharmacological effects of ethanol in the central nervous system.

In a second study, the effect of saffron extract on long-term potentiation of evoked potential in the dentate gyrus was investigated in anesthetized rats (Sugiura *et al.* 1995a). Interestingly, saffron was found to antagonize the long-term potentiation-blocking action of ethanol, at doses comparable to those which antagonized the memory-impairing effect of ethanol. The authors concluded that their results provide direct evidence of saffron extract's specific antagonizing action against ethanol, although they did not further clarify the underlying mechanism(s) of this effect.

A number of related studies have also been carried out with compounds which are known to be significant components of saffron extracts. In general, these have indicated that one compound in particular, crocin, is the most active as an ethanol antagonist and hence may be responsible for this activity of saffron extract (Morimoto *et al.* 1994, Sugiura *et al.* 1994, 1995a,b,c). These works have suggested that crocin may be useful as a pharmacological tool for studying the action of ethanol.

Saffron extract has also been shown to possess antitumor activity (that is, an inhibitory effect on tumor growth) and anticarcinogenic activity (that is, an inhibitory effect on the induction of cancer by carcinogens). The various studies which have been carried out on this effect of saffron are listed in Table 10.2.

The first report of the antitumor effect of saffron extract was published in 1991 (Nair *et al.* 1991a). This study showed that in mice, oral administration of saffron

Table 10.2 Antitumor effects of saffron and its components *in vivo* and *in vitro*

Agents	Tumor System	References
saffron	mouse S-180, EAC, DLA – *in vivo*	Nair *et al.* 1991a,b, 1992, 1994
saffron	mouse S-180, DLA – *in vivo*	Salomi *et al.* 1990, 1991a,b
saffron	mouse S-180, EAC, DLA, P388, osteosarcoma, ovarian sarcoma, fibrosarcoma – *in vitro*	Nair *et al.* 1991a,b, 1994, Salomi *et al.* 1990, 1991a,b
saffron	human HeLa, A549, VA13 – *in vitro*	Abdullaev *et al.* 1992a,b, 1996
crocetin	HL-60, K562 – *in vitro*	Tarantilis *et al.* 1992, 1994, Abdullaev 1994, Morjani *et al.* 1990, 1993
diimethyl crocetin	HL 60, K 562 – *in vitro*	Tarantilis *et al.* 1992, 1994, Morjani *et al.* 1990, 1993
crocin	HL 60, K 562 – *in vitro*	Morjani *et al.* 1990, Tarantilis *et al.* 1992, 1994
β-carotene	K 562 – *in vitro*	Morjani *et al.* 1990, Manfait *et al.* 1991

extract induces a dose-dependent inhibition of the intraperitoneal growth in mice of ascites tumors derived from sarcoma-180 (S-180), Erlich ascites carcinoma (EAC) and Dalton's lymphoma ascites (DLA) cells. Tumor-bearing mice which received 200 mg extract per kg body weight had significantly longer (two- to threefold) life spans than untreated tumor-bearing animals. Results of hematological and biochemical studies suggested that administration of this dose of extract to animals results in no overt toxicity (the LD_{50} was found to be 600 mg/kg). In a subsequent study (Nair *et al.* 1994), these authors found that oral administration of saffron extract in mice significantly inhibits the growth of solid tumors derived from DLA and S-180 cells, but does not affect the growth of solid tumors derived from EAC cells. They observed an elevation in the levels of β-carotene and vitamin A in the serum of the animals receiving saffron, and suggested this as a possible mechanism for the antitumor effect. In an interesting study, Nair *et al.* (1992) examined the efficacy of an alternative route of delivery: liposome-encapsulated saffron extract was injected intraperitoneally and the effect on tumor growth was examined. The authors concluded that liposome encapsulation enhances the antitumor activity of the extract towards several solid tumors, including the EAC tumor which was insensitive to orally administered extract. This enhancement in antitumor activity could be due to site-directed drug delivery or to carrier-mediated increased drug solubility (Nair *et al.* 1992).

Salomi *et al.* (1990, 1991a,b) examined the effect of saffron extract on the chemical induction of cancer in mice. They observed significant anti-carcinogenic activity of topically applied extract against dimethylbenz[a]anthracene-induced papillomas and of orally administered extract against methylcholanthrene-induced sarcomas. Of particular interest are several studies which have suggested the possible application of saffron extract in combination with "standard" chemotherapeutic agents to decrease their toxicity. Thus, Nair *et al.* (1991b) demonstrated that treatment with saffron extract prolongs the life span of cisplatin-treated mice twofold. Moreover the extract partially prevented the decrease in body weight, hemoglobin levels and leukocyte counts caused by cisplatin. Similarly, Salomi *et al.* (1991b) reported that saffron extract increases the life span of mice treated with chronic lethal doses of cyclophosphamide.

Studies *In Vitro*

A number of studies have demonstrated a cytotoxic effect of saffron extract on tumor cells *in vitro*. When trypan-blue dye exclusion was employed as a criterion of cell viability, the LD_{50} of the extract was found to range from approximately 7 µg/ml to 30 µg/ml, depending upon the type of tumor cells (Salomi *et al.* 1990, Nair *et al.* 1991a, 1994). Interestingly, there was no significant effect on normal mouse spleen cells, even at higher extract concentrations (Nair *et al.* 1991a). Other studies, utilizing colony formation as a measure of cell viability, showed that pretreatment of several types of tumor cells with saffron extract results in a dose-dependent decrease in their ability to form colonies, but has little or no comparable effect on normal cells.

Other studies have focused on the effects of saffron on various biochemical properties and processes of cells in culture. Exposure of tumor cells to saffron extract results in the inhibition of cellular nucleic acid synthesis (Abdullaev and Frenkel 1992a,b). Saffron has been shown to stimulate or support non-specific proliferation of immature and mature lymphocytes *in vitro* (Nair *et al.* 1992). The observed saffron-induced elevation in the intracellular levels of reduced glutathione and glutathione-related enzymes has suggested a possible antioxidant activity of saffron comparable to that of β-carotene (Nair *et al.* 1992).

A recent study (Abdullaev and Gonzalez de Mejia 1996) examined possible interactions between saffron and selenite, a compound with known anticarcinogenic activity (Combs and Combs 1986). Treatment of tumor cells with saffron in combination with selenite caused more effective inhibition of colony formation and nucleic acid synthesis relative to the effects of these agents alone. Treatment of tumor cells with saffron resulted in an increase in the level of intracellular sulfhydryl compounds (Nair *et al.* 1991a, Abdullaev and Gonzalez de Mejia 1996). Since the potency of selenite is known to correlate with the level of sulfhydryl compounds in the cell (Abdullaev *et al.* 1992c), this could explain the potentiation of selenite cytotoxicity by saffron.

In addition to these studies on the effects of saffron extract, there have been several investigations of the effects of some of its known components *in vitro*. It was demonstrated that the natural antioxidant 3, 8-dihydroxy-1-methyl antraquinone-2-carboxylic acid is present in saffron, and that it exhibits higher antioxidant activity than vitamin E in inhibiting the oxidation of linoleic acid (Isa 1992). Morjani *et al.* (1990) described the effects of the natural carotenoids crocin and its derivative dimethylcrocetin on K562 tumor cells. They reported that incubation with these compounds for 3 days results in significant inhibition of cell growth and differentiation. Tarantilis *et al.* (1992, 1994) investigated the potency of a variety of natural and semi-synthetic carotenoids, in comparison to that of retinoic acid. These compounds were highly effective in inhibiting the proliferation of HL-60 leukemic cells, as well as in inducing differentiation. The authors suggested that although the carotenoids are somewhat less potent than retinoic acid, they are expected to be less toxic and hence could prove useful in cancer chemotherapy.

To investigate whether the effects of saffron on cell proliferation can be accounted for by the effects of some of its major components, Escribano *et al.* (1996) compared the inhibitory effects of saffron extract to those of crocin, crocetin, picrocrocin and safranal. They concluded that the growth-inhibitory activity detected in total saffron extracts is mostly due to crocin. In particular, they found that crocetin exhibits very little cytotoxicity (as also reported by Abdullaev 1994 and Abdullaev and Gonzalez de Mejia 1996). This result suggests that sugars play a key role in the cytotoxic effect of crocin, since crocetin is its de-glycosylated derivative. Nevertheless, crocetin has been shown to inhibit cellular nucleic acid and protein synthesis (Abdullaev 1994), suggesting that it may also play a role in saffron cytotoxicity. In kinetic studies, Escribano *et al.* (1996) demonstrated that safranal has a more rapid effect than picrocrocin or crocin. This may reflect a better diffusion of safranal through the cell membrane due to its apolar nature and low molecular weight. They also described a number of morphological changes which are induced by crocin, including vacuolated areas, size reduction and condensed nuclei. These morphological changes might reflect the metabolic alterations which have been previously demonstrated at the molecular level in cells treated with saffron extract (Abdullaev and Frenkel 1992a,b).

Possible Mechanisms of Action

It is now generally accepted that cancer can be prevented by a variety of synthetic and naturally occurring compounds. Despite a large body of experimental and epidemiological evidence, the mechanism of action of most of these chemopreventive agents remains poorly understood. It should be noted that the complicated chemical composition of extracts of natural products makes it particularly difficult to determine the exact mechanism of their antitumor effects. Thus, it is perhaps not surprising that in spite of the recent evidence that saffron can have an inhibitory effect on experimental tumorigenesis and chemical carcinogenesis, the mechanism(s) of these effects remains unclear.

One general mechanism which has been proposed for the chemical prevention of tumorigenesis is a cytotoxic effect on tumor cells which prevents their proliferation and thus prevents the appearance of a tumor from the original transformed cell(s). Several of the studies described above have in fact demonstrated an inhibitory effect of saffron extract on cell proliferation. One of the most consistently observed cellular biochemical effects of saffron is its inhibitory effect on cellular DNA and RNA synthesis (Abdullaev and Frenkel 1992a,b, Nair *et al.* 1991a). It should be noted that in contrast to many cytotoxic agents, saffron extract has no significant inhibitory effect on cellular protein synthesis (Abdullaev and Frenkel 1992a,b). Of particular interest is the observation that saffron extract inhibited DNA and RNA synthesis in malignant human cells (irrespective of whether they originated from a tumor or from the transformation of normal cells *in vitro*) but had no detectable inhibitory effect on synthesis in non-malignant human cells (Abdullaev and Frenkel 1992a,b).

Taken as a whole, these findings suggest that the inhibitory effect of saffron on nucleic acid synthesis could represent a biochemical basis for its inhibitory effect on tumor-cell proliferation. The question has thus arisen of whether the inhibition of nucleic acid synthesis is in fact a direct effect of saffron, or the result of some other primary effect on the cell. This has been investigated by examining the effect of saffron on DNA and RNA synthesis in a cell-free system (isolated nuclei) (Abdullaev and Frenkel 1992a). The finding that saffron extract had no effect on the synthesis of DNA and RNA in isolated nuclei supports the conclusion that the inhibitory effect of saffron on cellular nucleic acid synthesis is probably not due to a direct effect on the synthetic reactions.

Several mechanisms have been proposed for the antitumor effect of the carotenoid constituents of saffron. The observation (Nair *et al.* 1992) that the antitumor effect of the extract could be demonstrated only when the drug was given orally but not when given intraperitoneally led to the hypothesis that prior metabolism of the active component/s may be required for its/their antitumor activity. Specifically, crocin is suggested to exert its antitumor effect via its metabolic conversion to a retinoid.

A second proposed mechanism for the antitumor action of carotenoids is based upon the widely accepted hypothesis that these compounds function as inhibitors of free radical chain reactions (Bruce 1983, Burton and Ingold 1984). Most carotenoids are lipid-soluble and thus might be expected to act as membrane-associated high-efficiency free-radical scavengers (Burton and Ingold 1984). This mechanism, involving the radical-trapping potential of carotenoids, has received support from computational molecular modeling studies (Neidle and Jenkins 1991, Martin 1991). A third mechanism involves the interaction of carotenoids with topoisomerase II, an enzyme involved in cellular DNA replication (Morjani *et al.* 1993). This idea is supported by the nuclear localization of some carotenoids (Manfait *et al.* 1991), as well as by their inhibitory effects on cellular DNA synthesis (see above). A fourth suggested mechanism is that the cytotoxic effect of crocin is mediated via apoptosis (Wyllie 1992).

CONCLUSION

As a result of a variety of recent studies, there is now convincing evidence for the biological activity of saffron and its constituents. These scientific findings, together with the body of anecdotal evidence for its therapeutic activity against a number of diseases, have provided strong indications that saffron and/or its components may prove to be useful agents in modern medicine. Future scientific investigations will undoubtedly focus on examining this possibility in appropriate animal models of human diseases. Additional studies will also be required to gain further understanding of the mechanism(s) of the biological effects of saffron. Such studies will undoubtedly uncover new biological activities and new potential applications. An important aspect of this will be the continuation and extension of current investigations on the identification and characterization of the biologically active compounds in saffron, and the definition of their modes of action at the molecular level.

One of the biological activities of saffron which has the greatest potential medical applicability is its ability to inhibit carcinogenesis. As described above, recent studies have shown that saffron extract possesses antitumor activity against transplanted tumors and anticarcinogenic activity against chemically induced carcinogenesis *in vivo* as well as cytotoxic effects on tumor-derived cells *in vitro*. Furthermore, the levels of saffron extract which were active in these experiments were nontoxic. These findings have raised the possibility that natural saffron and/or some of its constituents might be used as alternative antitumor or anticarcinogenic agents, either alone or in combination with synthetic substances having anticancer activity. Further investigation into the mechanisms of action of saffron extract will be important in this area as well. Once greater insight is achieved at the cellular and biochemical level, it should be possible to better assess which other agents are likely to act together with saffron in a synergistic manner. It should also be possible to better predict which types of protocols (e.g. chemopreventive or chemotherapeutic) are most likely to be successful, both in animal models and ultimately in human disease.

Natural plant extracts in general have proven to be an important source of antitumor agents, and compounds extracted from plants still provide some of the most original and promising approaches for discovering new drugs. The studies described here provide initial indications that this is likely to prove true for saffron as well. It is reasonable to presume that at this point, we have only begun to scratch the surface of the potential applications of saffron in human health and disease.

ACKNOWLEDGEMENTS

The authors wish to thank Drs. Morimoto, Nair, Ni, Shoyama and Tarantilis, for providing their papers and helpful assistance.

REFERENCES

Abdullaev, F.I. (1993) Biological effects of saffron. *BioFactors*, **4**, 83–86.
Abdullaev, F.I. (1994) Inhibitory effect of crocetin on intracellular nucleic acid synthesis in malignant cells. *Toxicology Lett*, **70**, 243–251.
Abdullaev, F.I. and Frenkel, G.D. (1992a) Effect of saffron on cell colony formation and cellular nucleic acid and protein synthesis. *BioFactors*, **3**, 201–204.

Abdullaev, F.I. and Frenkel, G.D. (1992b) The effect of saffron on intracellular DNA, RNA and protein synthesis in malignant and non-malignant human cells. *BioFactors*, **4**, 38–41.

Abdullaev, F.I., MacVicar, C. and Frenkel, G.D. (1992c) Inhibition by selenium of DNA and RNA synthesis in normal and malignant human cells *in vitro*. *Cancer Lett*, **65**, 43–49.

Abdullaev, F.I. and Gonzalez de Mejia, E. (1996) Inhibition of colony formation of Hela cells by naturally occurring and synthetic agents. *BioFactors*, **6**, 1–6.

Bruce, N.A. (1983) Dietary carcinogens and anticarcinogens. Oxygen radicals and degenerative diseases. *Science*, **221**, 1256–1264.

Burton, G.W. and Ingold, K.U. (1984) β-carotene: An unusual type of lipid antioxidant. *Science*, **224**, 569–573.

Castellar, M.R., Montijano, H., Manjon, A. and Iborra, J.L. (1993) Preparative high-performance liquid chromatographic purification of saffron secondary metabolites. *J. Chromatography*, **648**, 187–190.

Combs G.F. and Combs, S.B. (1986) The Role of *Selenium in Nutrition*. Academic Press, Orlando, FL.

Damirov, I.A., Prilipko, L.I., Shukurov, D.Z. and Kerimov, J.B. (1988) *Remedy Plants of Azerbaijan*. Maarif, Baku, Azerbaijan, pp. 90–93.

Escribano, J., Alonso, G.L., Coca-Prados, M. and Fernàndez, J.A. (1996) Crocin, safranal and picrocrocin from saffron (*Crocus sativus* L.) inhibit the growth of human cancer cells *in vitro*. *Cancer Letters*, **100**, 23–30.

Gainer, J.L. and Jones, J.R. (1975) The use of crocetin in experimental atherosclerosis. *Experientia*, **31**, 548–549.

Isa, T. (1992) Antioxidative property of the anthraquinone-pigment from the cultured cells of saffron, and enzymatic comparison between some cultured cells. *Shokubutsu Soshiki Baiyo*, **9**, 51–53.

Kasumov, F.J. (1970) *The Extract of Saffron Flowers*. GosPlan Press, Baku, Azerbaijan.

Liakopoulou-Kyriakides, M. and Skubas, A.I. (1990) Characterization of the platelet aggregation inducer and inhibitor isolated from *Crocus sativus*. *Biochemistry International*, **22**, 103–110.

Manfait, M., Morjani, H., Efremov, R., Angibousi, J.F., Polissiou, M. and Nabiev, F. (1991) High sensitive detection of intracellular carotenoids in single living cancer cells as probed by surface-enhanced Raman spectroscopy. In R. E. Hester and R. B. Giring (eds.), *Spectroscopy of Biological Molecules*, Royal Society of Chemistry, UK, pp. 303–304.

Martin, Y.C. (1991) Computer-associated rational drug design. In: J. Langone (ed.), *Methods in Enzymology, Molecular Design and Modeling: Concepts and Applications*, Vol. **203** Part B: *Antibodies and Antigens, Nucleic Acids, Polysaccharides, and Drugs*, Academic Press, New York, pp. 587–613.

Miwa, T. (1954) Study on *Gardenia florida* L. (Fructus gardeniae) as a remedy for icterus. *Japanese J. Pharmacol*, **4**, 69–73.

Morimoto, S., Umezaki, Y., Shoyama, Y., Saito, H., Nishi, K. and Irino, N. (1994) Post-harvest degradation of carotenoid glucose esters in saffron. *Planta Med.*, **60**, 438–440.

Morjani, H., Tantilis, P., Polissiou, M. and Manfait, M. (1990) Growth inhibition and induction of erythroid differentiation activity by crocin, dimethylcrocetin and b-carotene on K562 tumor cells. *Anticancer Res.*, **10**, 1398–1406.

Morjani, H., Riou, J.F., Nabiev, Y., Lavelle, F. and Manfait, M. (1993) Molecular and cellular interaction between intoplicine, DNA and topoisomerase II studied by surface-enhanced Raman scattering spectroscopy. *Cancer Res.*, **53**, 4784–4790.

Nadkarni, K.M. (1976) *Crocus sativus, Nigella sativa*. In K.M. Nadkarni (ed.) *Indian Materia Medica.*, Popular Prakashan, Bombay, pp. 386–411.

Nair, S.C., Pannikar, B. and Panikkar, K.R. (1991a) Antitumor activity of saffron (*Crocus sativus*). *Cancer Lett.*, **57**, 109–114.

Nair, S.C., Salomi, M.J., Panikkar, B. and Panikkar, K.R. (1991b) Modulatory effects of the extracts of saffron and *Nigella sativa* against cisplatinum induced toxicity in mice. *J. Ethnopharmacol.*, **31**, 75–83.

Nair, S.C., Salomi, M.J., Varghese, C.D., Panikkar, B. and Panikkar, K.R. (1992) Effect of saffron on thymocyte proliferation, intracellular glutathione levels and its antitumor activity. *BioFactors*, **4**, 51–54.

Nair, S.C., Varghese, C.D., Panikkar, K.R., Kurumboor, S.K. and Parathod, R.K. (1994) Effects of saffron on vitamin A levels and its antitumor activity on growth of solid tumors in mice. *Int. J. Pharmacog.*, **32**, 105–114.

Neidle, S. and Jenkins, T.C. (1991) Molecular modeling to study DNA interaction by antitumor drugs. In J. Langone, (ed.), *Methods in Enzymology, Molecular Design and Modeling: Concepts and Applications*, Vol. **203** Part B: *Antibodies and Antigens, Nucleic acids, Polysaccharides, and Drugs*, Academic Press, New York, pp. 433–458.

Ni, X. (1992) Research progress on the saffron (*Crocus sativus*). *Zhongcaoyao*, **23**, 100–107.

Pfander, H. and Witwer, F. (1975) Untersuchungen Zur Carotinoid-Zusammensetzung in Safran. *Helv. Chim. Acta*, **58**, 1608–1620.

Salomi, M.J., Nair, S.C. and Panikkar, K.R. (1990) Inhibitory effects of *Nigella sativa* and saffron (*Crocus sativus*) on chemical carcinogenesis in mice and its non-mutagenic activity. *Proc. Ker. Sci. Congress*, **3**, 125.

Salomi, M.J., Nair, S.C. and Panikkar, K.R. (1991a) Inhibitory effects of *Nigella sativa* and saffron (*Crocus sativus*) on chemical carcinogenesis in mice. *Nutrition and Cancer*, **16**, 67–72.

Salomi, M.J., Nair, S.C. and Panikkar, K.R. (1991b) Inhibitory effects of *Nigella sativa* and saffron (*Crocus sativus*) on chemical carcinogenesis in mice and its non-mutagenic activity. *Proc. Ker. Sci. Congress*, **3**, 344–345.

Sugiura, M., Shoyama, Y., Saito, H. and Abe, K. (1994) Crocin (crocetin di-gentiobiose ester) prevents the inhibitory effect of ethanol on long-term potentiation in the dentate gyrus *in vivo*. *J. Pharmacol. Exp. Ther.*, **271**, 703–707.

Sugiura, M., Saito, H. and Abe, K. (1995a) Ethanol extract of *Crocus sativus* L. antagonizes the inhibitory action of ethanol on hippocampal long-term potentiation *in vivo*. *Phytotherapy Res.*, **9**, 100–104.

Sugiura, M., Shoyama, Y., Saito, H. and Abe, K. (1995b) The effects of ethanol and crocin on the induction of long-term potentiation in the CA1 region of rat hippocampal slices. *Jap. J. Pharmacol.*, **67**, 395–397.

Sugiura, M., Shoyama, Y., Saito, H. and Nishiyama, N. (1995c) Crocin improves the ethanol-induced impairment of learning behaviors of mice in passive avoidance tasks. *Proceedings of the Japan Academy*, **71**, ser. B, 10, 319–324.

Sujata, V., Ravishankar, G.A. and Venkataraman, L.V. (1992) Methods for the analysis of the saffron metabolites crocin, crocetins, picrocrocin and safranal for the determination of the quality for the spice using thin-layer chromatography, high-performance liquid chromatography and gas chromatography. *J. Chromatography*, **624**, 497–502.

Suzhou New Medical College (1977) *Dictionary of Traditional Chinese Medicine (Zhong Yao Da Zi Dian)* Vol. 2. Shanghai People's Publication House, Shanghai, pp. 2622–2623.

Tarantilis, P.A., Polissiou, M., Morjani, H., Avot, P., Bei Jebbar, A. and Manfait, M. (1992) Anticancer activity and structure of retinoic acid and carotenoids of *Crocus sativus* L. on HL60 cells. *Anticancer Res.*, **12**, 1989–1992.

Tarantilis, P.A., Morjani, H., Polissiou, M. and Manfait, M. (1994) Inhibition of growth and induction of differentiation of promyelocytic leukemia (HL-60) by carotenoids from *Crocus sativus* L. *Anticancer Res.*, **14**, 1913–1918.

Wyllie, A.H. (1992) Apoptosis and the regulation of cell numbers in normal and neoplastic tissues: an overview. *Cancer Metastasis Rev.*, **11**, 95–103.

Zhang, Y., Shoyama, Y., Sugiura, M. and Saito, H. (1994) Effects of *Crocus sativus* L. on the ethanol-induced impairment of passive avoidance performances in mice. *Biol. Pharm. Bull.*, **17**, 217–221.

Zhou, Q., Sun, Y. and Zhang, X. (1987) Saffron, *Crocus sativus* L. *J. Traditional Chinese Med.*, **28**, 59–61.

11. MECHANIZED SAFFRON CULTIVATION, INCLUDING HARVESTING

PIER FRANCESCO GALIGANI and FRANCESCO GARBATI PEGNA

Dipartimento di Ingegneria Agraria e Forestale, Università degli Studi di Firenze, P. le delle Cascine, 15-50144 Firenze - I, Italia

ABSTRACT   Saffron is quite a difficult crop to mechanize since the plant is small and delicate and in some phases, such as harvest or corm gathering and planting, rather complicated to handle. However, some common tools, created mainly for other cultivated plants, can be successfully adapted for use in most cultural phases: some of the tools, tested in specific trials, are described and evaluated together with others particularly designed for the cultivation of saffron.

Saffron cultivation is not highly mechanized, even in this day and age: although it requires high labour input during the most important growing phases, there are no machines capable of totally mechanizing this crop, and research up to now has always tried to adapt existing machinery to each individual phase of its cultivation, rather than design specific machines.

The reasons for this can be attributed to the delicacy of certain operations and to the marginal nature of this type of cultivation. In fact, corms are very delicate and need to be handled with care; they also vary considerably in size, and this makes mechanical handling difficult. Moreover production is limited to a small anatomical part of the plant which is difficult to reach and to separate out.

Although it is certainly possible to overcome these difficulties – agricultural mechanization has tackled far more complex problems – the low profitability of saffron has placed it in such a marginal position, especially in countries where labour costs are higher and where mechanization would therefore be more useful, that it is not economically viable to invest resources in searching for valid solutions. In fact the limited amount of land generally devoted to this crop does not encourage manufacturers to take an interest in the sector as investments are unlikely to be repaid by the sale of their products.

Nonetheless, between 1980 and 1985, the problem of mechanized saffron cultivation was faced in Italy, within the framework of research into medicinal plants funded by the Ministry of Agriculture and Forestry (M.A.F.). This resulted in the identification of certain possible working hypotheses, without, however, yielding any concrete results in the most complex phases of the growing cycle, namely pistil collection and stamen separation (M.A.F. 1981, 1982, 1983).

This study (to which reference will continually be made hereafter owing to the scarcity of other sources of experimental data on the subject), while stressing the difficulties involved in carrying out most of the operations that are characteristic of saffron growing, has nonetheless produced some results. It is interesting to note that adapting machines already available for other crops can lead to considerable savings in labour for some of the more complex operations. Experiments on saffron, including mechanical cultivation, are still under way in Italy, including the area of San Gimignano in the province of Siena, where this crop, once very important for the local economy (Landi 1996), is now being reintroduced.

A series of trials enabled a determination of the workload involved in performing each of the growing operations as they are done in Italy, using the tested machinery (Adamo *et al.* 1987). Considerably varied data emerged from this study, due principally to the small size of the plots used for saffron growing and their variable physical and agronomic characteristics. However, although expressed in mean values, these data (Table 11.1) are a useful reference for those who wish to evaluate the efficacy of the various options for mechanization.

In summary, saffron growing can, in part, be performed using commercially available agricultural machinery as is or following some simple adjustments, but certain operations must still be done by hand. This is especially true in the more delicate phases, for which specially built equipment could be used to aid in the operations, but for which total mechanization is difficult to envisage. This problem, however, should be more specifically analysed, focusing on the state of the art and providing a starting point for further improvements: we therefore present a brief description of the various phases of saffron growing, with a few remarks on the possibilities for mechanization, on the basis of experiments conducted within the aforementioned M.A.F. project.

PREPARATION FOR PLANTING

Preparation of the land for saffron planting generally involves tilling the soil to a depth of approximately 300 mm and then improving it. This operation is not difficult to mechanize since tilling and improving can be performed using the usual equipment available on most farm estates, while in the case of the small family farms which do not take part in large-scale farming activities, a suitably equipped walking tractor is sufficient: a 10-kW walking tractor with a ridge plough and hoeing machine enables an area of 1000 m^2 to be worked in about 2 h and to be improved in about 1 h.

When preparing the soil it is, however, very important to avoid water stagnation, which is very dangerous for this crop, so the ground is often prepared in ridges to help the water drain off. The ridges can be made in various ways, but the best implement is any kind of ridger which, with a little effort, can be combined with the machines used for planting vegetative material. In the trials conducted in Italy with a ridger alone, or in combination with a potato planter (described further on), it was possible to make ridges about 150 mm high and 1 m wide separated by furrows of about 300 mm (Galigani 1987). However, the rate of this type of machine varies according to the conditions of use, fluctuating between 2 and 10 h/1000 m^2 (Amato *et al.* 1989).

Basic fertilization can also be performed using the usual commercially available equipment, although in view of the small size of the plots it is often done by hand.

Table 11.1 Time required and methods of performing the principal growing operations at Navelli (h/1000 m²)

Growing Operation	Method Used	Location	IX	X	XI	XII	I	II	III	IV	V	VI	VII	VIII	TOT	%
Ploughing	2-wheel tract.	field	1.3												1.3	0.3
Manuring	Spreading	field	3.6												3.6	0.8
Hoeing	2-wheel tract.	field											0.8		2.4	0.5
Transplanting	manual	field							0.8		0.8			33.3	33.3	6.9
Ridging	manual	field	6.7												6.7	1.4
Harvesting	manual	field		52.0	48.0										100.0	20.8
Separating	manual	farmhouse		80.0	160.0										240.0	50.0
Mowing	manual	field									5.3				5.3	1.1
Corm picking	manual	field							0.8		6.1		0.8	33.3	33.3	6.9
Selection	manual	farmhouse												54.1	54.1	11.3
Total			11.6	132.0	208.0	0	0	0	0.8	0	1.3	0	0.2	120.7	480.0	100.0
%			2.4	27.5	43.3	0	0	0	0.2	0				25.0	100.0	

For annual fertilization, conveyor feed distributors can be used to avoid burning the leaves, as may happen when applying urea during the vegetative period (Galigani 1987).

PLANTING

The saffron corm poses many problems with regard to the mechanization of planting, because the vegetative material is small and delicate and requires regular and correctly oriented placement.

The most suitable machines for this operation are onion planters, which nevertheless need to be slightly modified, in particular to adapt them to the size of the corms. The machinery we tested is of the type carried by the three-point linkage on any kind of tractor, even a light two-wheel-drive tractor. It has a 1.5-m working width and consists of a bulb hopper and a series of scoop wheels which lift the corms out of the hopper and drop them into a funnel. This funnel has a furrowing unit at the bottom and a covering unit at the rear end. The planting distances along the row can be varied between 20 and 120 mm, by altering the transmission ratios connecting the scoop wheels to one of the rear wheels. The machine weighs about 400 kg when empty.

An important drawback of this type of machine is that it deposits the corms in the ground without respecting their polarity, so that some corms are planted leaning away from their vertical axis and others are even upside down: trials to assess the effect of unnatural corm positioning (Table 11.2) have shown that a leaning position causes a delay in sprouting but increased flower production, whereas the upside-down position causes both a delay in sprouting and a marked decrease in production[1] (Galigani 1982). The working time with this planter per 1000 m^2 arranged in ridges (four rows per ridge with a total of 55,000 corms) is 5 h, vs. over 100 h for manual planting (Figure 11.1).

Another type of machine that can be adapted to saffron planting is the potato planter: the corms are placed by hand in the scoops, which are moved in horizontal rows by a wheel resting on the ground, by means of chains. The chains are lowered into the ground to deposit the corms in the furrows opened by a furrowing implement located at the front; at the rear, two discs close the furrow.

In the trials, however, only two rows could be planted per ridge, owing to the size of the implements and to the structure, which is intended for only two

Table 11.2 Influence of planting position of corms on the time of emergence and on production

	Normal	*Upside Down*	*In Between*
Emergence after 8 days	86.5%	33.3%	43.2%
Total number of shoots	control	−28.58%	+10.70

---

[1] Other authors have reported a 60% reduction in blossom of tilted corms.

Figure 11.1

operators. Moreover it was not possible to adjust the distance along the row to less than 150 mm. Overall, this machine was found to give a lower yield than the onion planter but to provide better control over corm orientation (Galigani 1987, Galigani and Adamo 1987, Tammaro 1990).

The potato planter can also be combined with a ridger to prepare and plant in a single operation. This combination produces satisfactory results, reducing the working times (ridging and planting) per 1000 m^2 to 24 h (8 h/1000 m^2 for a machine with three operators) (Amato *et al.* 1989).

With regard to trials in Italy, during the 1980s tests were carried out involving burying zinc-mesh cages with a U cross-section (1000 × 80 × 60 mm) containing the corms: this was meant to facilitate the subsequent extraction of the corms from the earth at the end of the cycle (Figure 11.2). This solution was slightly better than the traditional method in terms of the time involved, but the cages were easily damaged, carrying the consequent risk of a considerable increase in costs (Galigani 1987); each cage can be expected to last for 3 years. The working times involved for an area of 1000 m^2 are reported in Table 11.3. Although the cage system seems to provide the operator with more comfort, it has not met expectations due to the tendency of the cages to warp and of the corms to slide about inside the cages during removal, consequently causing variations in planting density.

WEEDING AND CULTIVATION

The problem of weeds in the first year of cultivation is practically non-existent as blossoms sprout a reasonably short time after having cleared the terrain for the pre-planting stage.

Figure 11.2

In perennial crops with no ridging, normal hoeing operations can be executed at the beginning of the second year, the work being finished manually or with the aid of a rotating hoe applied to a back-mounted scrub-clearing machine.[2] More often, however, repeated milling or superficial harrowing are effected with normal tools, even those transported on light tractors. If the terrain has been ridged, on the other hand, it is necessary to use two-wheel tractors, taking care not to till at depths below 20 or 30 mm because of the tendency of new corms to grow increasingly close to the surface. Mulching also gives good results, especially using wood chips or sawdust.

Alternatively, spring or summer mowing can be carried out to eliminate the infesting weeds together with any residual crop leaves, which in any case would be lost in the summer stagnation. The mowed vegetation can be made into hay and used as animal fodder. Otherwise flaming can be applied, using suitable commercially available equipment: these machines can be carried on the back or on a hand-propelled trolley. The results with the latter technique are good as far as

---

[2] This is a small cutter with vertical axis, made by adapting a scrub-clearing machine with a combustion engine, and carried on the back. The blade is replaced with an eccentric mass disk and a counterdisk equipped with points [Z].

Table 11.3 Workload required for 1000 m² using the cage system

Operation	h
Ditching	34
Cage building*	75
Positioning of the corms	20
Positioning of the cages	6

*total time for building = 225 h, duration 3 years: workload for each year = 75 h.

young weeds are concerned, if weeding is performed during the hottest, driest hours of the day. Otherwise plants must be kept in contact with the flame as long as possible, because of a sizeable reduction in the operative capabilities of the system (Galigani 1987, Galigani and Adamo 1987).

In terms of production, no significant differences between mulching, flaming and hoeing have been noted in the tests, although of the three, mulching is the least labour-intensive technique (Landi 1996). Chemical weeding does not seem advisable for this type of crop because it may pollute the product.

BLOSSOM COLLECTION

Saffron flowers are normally hand-picked in the early morning hours, when the corolla is still closed, because of the better quality obtained. Average production per hectare fluctuates around 5000 kg. Work is lengthy and hard-going, as the picker has to assume a very uncomfortable, stooping position and must cut each blossom at the base of the corolla, at most using his thumbnail. Mean hourly productivity per person is estimated at between 8 and 16 kg (2000–4000 flowers).

The simplest solution would seem to be the use of mowing or grass-cutting machines which have been especially calibrated to cut the blossoms very low down, collecting up the mowed vegetation and then separating the flowers from the leaves. However, it has not so far been possible to do so on account of saffron's graded flowering times and because cutting off the leaves adversely affects the corm's future chances of development (Tammaro 1990).

For these reasons, harvesting is considered one of the most difficult operations to mechanise, given the delicacy and precision required. Indeed, the sensitivity of this operation does not permit the use of any of today's commercially available machinery, and custom-designing and fabricating one seems a very complex exercise. Furthermore, as saffron is a marginal crop, it is unlikely that the effort required to solve the problem would be economically worthwhile. The only direction to follow thus seems to be to obtain facilitating tools by adapting equipment which is already commercially available.

Some tests have been performed by adapting a vacuum machine for collecting olives or dry leaves from the ground. This machine can be back-mounted and consists of a 40 cc engine driving a suction fan linked to a metal pipe. Horizontal scissors controlled manually from above are applied to the end of the pipe (Figure 11.3). Two small wheels maintain the height of flower cutting above the ground at a constant level (about 40 mm) (M.A.F. 1981).

Figure 11.3

This device has proven valid operationally, but not economically because of the low yield and the high percentage of product which is passed over: cutting the blossoms at ground height does not allow the picking of the whole style. In any case, leaves and other impurities get mixed in. Besides, picking in the early hours means the inconvenience of closed flowers which are more aerodynamic and not so easy to suck up. Other disadvantages arise in the next stage of crop sorting because of the higher amount of impurities (M.A.F. 1983). The device described here consequently turns out to be useful for improving the operator's work stance rather than his productivity (M.A.F. 1982, Galigani 1987).

However, mechanized harvesting using this method seems to have a negative effect on the colouring power of the spice, whereas the potency of the bitter flavour appears to be stimulated to some degree, probably because the motion of sucking up under a draught accelerates the drying process, encouraging the transformation of picrocrocin into saffranal, and also because mechanical cutting eliminates some of the basal portion of the stigma.

A simpler solution seems to be that of continuing to cut the blossoms by hand, but facilitating their placement into containers by using the described sucking system. However, the scissors would have to be removed and the machine placed in a position which allows closed flowers to be sucked up as well. This would seem possible through the use of facilitating machines of the type used for picking asparagus or strawberries: the picker's stance would be improved and the harvesting bags brought nearer to ground level, thereby allowing the blossoms to be sucked up. A machine of this kind is under study at the moment at our department in Florence.

SEPARATION OF THE STIGMAS

Average hourly productivity in this activity per person fluctuates between 500 and 1500 blossoms, and the time needed to separate the flowers produced over 1000 m^2 (about 140,000) therefore varies from 93 to 280 h.

Sorting is always done by hand, even though attempts have been made to separate the styles from the stamens and petals by means of a wind tunnel consisting of a variable-section pipe which exposes the cut flowers to an air draught made up of various vortices. The styles remain, but they are sometimes attached to the perianth to which they are naturally joined in the basal section of the pistil. This also happens in flowers gathered by cutting and sucking up, as the two parts of the flower are still joined (despite the cut separating that part of the pistil attached to the base of the perianth from the free part) because of the high internal humidity of the blossom, due to harvesting in the early morning. Reduction in humidity by means of drying after harvesting does not resolve the problem, since the corolla curls and causes the pistils to cling together even more tenaciously (M.A.F. 1981, 1982, Galigani 1987).

In a simplified version of this appliance, the petals are separated from the stamens by a fan and then separated manually or by means of a flat or cylindrical iron screen, but this operation also needs to be completed by hand (Skrubis 1990). The use of vibrating boards has not proven suitable either, in separating the stigmas from the stamens and petals (Galigani and Adamo 1987).

DRYING

Drying is generally carried out on silk trays placed on shelves in a dark, stove-heated dryer for about 12 h. The time needed using this technique to dry the produce from 1000 m^2 (5–6 kg) is about 17 h. Alternatively, dehydration chambers have been tried out in laboratories where the crop is maintained at a temperature of 48°C for 3 h. The results are good (Skrubis 1990) in terms of time, but the use of electrical dryers seems to decrease the crop's organoleptic qualities (Tammaro 1990).

Once the drying stage is over, the crop may be ground into a powder. Electric coffee-grinders have proven useful for this operation (Tammaro 1990).

CORM GATHERING

The most common uprooting method is manual, by means of a hoe or small plough with a single or double ploughshare which, on turning over the soil, brings the corms to the surface to be hand-picked. To facilitate this second operation, ploughs with open mouldboards are often employed (Figure 11.4).

Otherwise bulb- or tuber-picking machines may be used. In either case, specific adaptations need to be made. For example, the use of common potato diggers is

Figure 11.4

possible, regulated so as to reduce the depth of digging and increase the workfront, in consideration of the more superficial collocation of the saffron corms in the ground and their very limited size as compared to potatoes. Even the vibrating grid at the back which sorts the produce that has been dug up must be modified by thickening the links (Galigani 1987, Galigani and Adamo 1987).

In trials carried out during the M.A.F. project cited earlier, a potato-picking machine drawn by a two-wheel-drive tractor weighing 270 kg was used with success (Figure 11.5). This machine consists of a neoprene frame with a front three-pointed ploughshare of semi-cylindrical shape followed by a rod-iron grid normally placed in the feed direction. The grid is hinged at one end and is made to swing by an eccentric moved by the power take-off. Work depth is regulated to 150–180 mm. Use of the original version of this machine has pinpointed the usefulness of the above modifications so as to have a three-pointed ploughshare and a longer vibrating grid with thicker links and more teeth when undersized corms are to be gathered. Either way, results have been satisfactory, especially in bare stony soil, and the loss of the smaller, hardly commercially viable, cormels does not affect the economy of the operation (Galigani 1982, M.A.F. 1982, Amato et al. 1989).

Figure 11.5

In trials carried out with the aforementioned cages, retrieval of a greater number of corms as compared to cultivation in free soil is possible, but the average size of the corms is inferior and work hours are not substantially reduced. When, on operating between the rows with a three-element cultivator, the cages have been uncovered, it is then necessary to finish the work by hand with a hoe. The cages are lifted manually and emptied onto sheets where the corms can be cleaned (M.A.F. 1983). This technique does not seem very viable economically, especially on account of the laborious operation and the high cost of the cages.

CONCLUSIONS

The mechanization of saffron cultivation presents various difficulties, linked mostly to the particular characteristics of the plant,[3] the localization of the crop and the marginality of its cultivation. However, some of the cultivation phases may make use of machines used in other, more frequently practised crop cultivation which, with simple adaptations (or at times even without), allow a considerable reduction in man hours.

Regarding other operations, and specifically the harvesting, sorting and processing of the stigmas,[4] mechanized methods still need to be invented, even though today we have at our disposal studies on the subject highlighting the main problems and indicating some possible solutions. Mechanizing these stages does, however,

---

[3] Note that working on a genetically sterile plant imposes a considerable obstacle to mechanisation.
[4] In saffron cultivation as presently carried out in Italy, 40% of labour is taken by stigmas' separation, 15% by blossom picking and 5% by toasting and packaging.

assume a certain degree of modernization in saffron cultivation, to move it from traditional agriculture on a small scale carried out by small farmers on marginal land, to a more dynamic agriculture where the cost of designing and making specific machines is paid for by the expansion of allotments, and economies of scale may compensate for the inevitable decline in care in the execution of each operation.

This course of action bodes well, however, because of the very high production value. Unfortunately, today this value involves costs that are just as high. Hence labour remuneration turns out to be equal to that of other, less valuable crops, consequently severely limiting the convenience of this cultivation.

REFERENCES

Adamo, A., Cozzi, M., Galigani, P.F., Vannucci, D. and Vieri, M. (1987) Fabbisogno di manodopera nelle operazioni colturali dello Zafferano, in *Atti, Convegno sulla coltivazione delle piante officinali, Trento 9–10 ottobre 1986,* ed. A. Bezzi pp. 451–452, Istituto Sperimentale per l'Assestamento Forestale e per l'Alpicoltura, Villazzano (Trento).

Amato, A., Amelotti, G., Bianchi, A., Galigani, P.F., Montorfano, P. and Zanzucchi, C. (1989) Zafferano, fonte di reddito alternativo per le zone svantaggiate. *Agricoltura*, 196, 101–128.

Galigani, P.F. (1982) Progetto Piante Officinali: Relazione dell'attività svolta dall'Unità Operativa dell' Istituto di Meccanica Agraria e Meccanizzazione della Facoltà di Agraria dell'Università di Firenze nel II anno di ricerca 1981–1982. Unpublished.

Galigani, P.F. (1987) La meccanizzazione delle colture di salvia, lavanda, zafferano e genziana, in *Atti, Convegno sulla coltivazione delle piante officinali, Trento 9–10 ottobre 1986,* ed. A. Bezzi, pp. 221-234, Istituto Sperimentale per l'Assestamento Forestale e per l'Alpicoltura, Villazzano (Trento).

Galigani, P.F. and Adamo, A. (1987) Le macchine per le officinali. *Terra & Vita*, 10, 62–7.

Landi, R. (1996) Relazione sull'attività svolta dall'Associazione "Il Croco" di S. Giminiano nell'ambito del programa di ricerche condotte con il contributo della Regione Toscana – 1994–1996.

M.A.F. (1981) Progetto piante officinali: stato della sperimentazione e risulati del primo anno di attività – Ministero dell'Agricoltura e delle Foreste – Istituto Sperimentale per l'Assestamento Fortestale e per l'Alpicolturta di Trento.

M.A.F. (1982) Progetto piante officinali: stato della sperimentazione e risulati del secondo anno di attività – Ministero dell'Agricoltura e delle Foreste – Istituto Sperimentale per l'Assestamento Fortestale e per l'Alpicolturta di Trento.

M.A.F. (1983) Progetto piante officinali: stato della sperimentazione e risulati del terzo anno di attività – Ministero dell'Agricoltura e delle Foreste – Istituto Sperimentale per l'Assestamento Fortestale e per l'Alpicolturta di Trento.

Skrubis, B. (1990) The cultivation in Greece of *Crocus sativus* L. *Proceedings of the International Conference on Saffron (Crocus sativus* L.), *L'Aquila (Italy) October, 27–29 1989,* eds. F. Tammaro and L. Marra, pp. 171–182, Università degli Studi dell'Aquila, Accademia Italiana della Cucina, L'Aquila.

Tammaro, F. (1990) *Crocus sativus* L. cv. Piano di Navelli – L'Aquila (L'Aquila saffron): environment, cultivation, morphometric characteristics, active principles, uses. *Proceedings of the International Conference on Saffron (Crocus Sativus* L.) *L'Aquila (Italy) October, 27–29 1989,* eds. F. Tammaro and L. Marra, pp. 47–98, Università degli Studi dell'Aquila, Accademia Italiana della Cucina, L'Aquila.

12. STERILITY AND PERSPECTIVES FOR GENETIC IMPROVEMENT OF *CROCUS SATIVUS* L.

GIUSEPPE CHICHIRICCÒ

Department of Environmental Sciences,
University of L'Aquila,
Via Vetoio, 67100 L'Aquila, Italy

ABSTRACT   In the saffron crocus, the developmental potential of the spore mother cells is limited by their triploid genome which causes meiotic abnormalities, followed by variations in sporogenesis and gametogenesis. As a result, abnormal gametophytes are generated. However, the reproductive system of the saffron crocus, like that of fertile *Crocus* species, supports interspecific crosses with related species. This potential cross-compatibility, together with *in vitro* methods which raise successful seed set, may open the door to breeding programmes for the genetic improvement of the saffron crocus.

INTRODUCTION

The reproductive cycle of angiosperms includes two generations, sporophytic (diploid) and gametophytic (haploid). The gametophytic generation is extremely reduced; the male (pollen) consists of a vegetative cell and two sperm cells, and the female (embryo sac) of seven cells, including the gametic cells (egg cell and central cell). The union of sperm cells with the female gametes (double fertilization) requires the development of a pollen tube to convey sperm cells through the pistil to the embryo sac. Pollen-tube development results from a continuous interaction with the transmitting tissue of the pistil; the interaction is controlled by genetic systems which prevent growth after either cross- or in-breeding (see De Nettancourt 1977). The transition from sporophytic to gametophytic generations occurs via the meiotic process. This comprises a series of coordinated developmental stages of the sporocyte, also correlated with the development of the surrounding sporangium tissues. Any developmental abnormality during meiosis may result in gametophytic sterility. A factor usually associated with abnormal meiosis is polyploidy. The pollen of polyploid plants, especially triploids, shows a variable degree of pollen sterility (see Carroll 1966). Female sterility is less known because it is difficult to evaluate.

This chapter reviews studies on the reproductive system of the saffron crocus (*Crocus sativus* L.), and its potential with respect to future perspectives for its genetic improvement.

KARYOLOGY

The saffron crocus genome consists of 24 chromosomes, morphologically grouped into a triplicate set of eight chromosomes each. Its karyotypic configuration shows few variations within and between European and Asiatic populations (Brighton 1977). Heteromorphism has been observed in populations from Majorca, Spain (Brighton 1977), and L'Aquila, Italy (Chichiriccò 1984), the former relative to two chromosomes and the latter to a single one. The meiotic paring of chromosomes in triplets, as established in Japanese (Karasawa 1933), Italian (Chichiriccò 1984) and Iranian (Ghaffari 1986) populations, asserts autotriploidy. The mean frequency of trivalents during metaphase I is 7.3 per cell, as evaluated in the Italian population.

The meiotic divisions are characterized by abnormalities typical of polyploids (Chichiriccò 1984, Ghaffari 1986). In meiosis I, irregular chromosome segregation arises from trivalent formation; either one or two chromosomes of each trivalent may go to one of the spindle poles to be included in the daughter nuclei. As a rule, the sharing of chromosomes between the two poles is rather unbalanced: it varies numerically from 8 to 15, and imbalance may be caused by lagging chromosomes such as univalent chromosomes. These either fail to reach the poles, or exhibit alternative segregation through precocious division of the chromatids. The genic unbalance of the nuclei increases through meiosis II, owing to laggards and extra-polar assortment of chromatids during anaphase II. As a result of this erratic chromosome assortment, meiosis culminates in abnormal cytokinesis producing a number of spores which differ from the standard quartet (Figure 12.1).

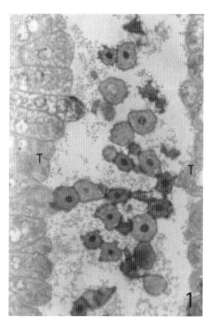

Figure 12.1 Longitudinal section of a saffron crocus anther showing tapetal cells (T) and microspores after release from tetrad callose (× 295).

MICROSPOROGENESIS AND POLLEN DEVELOPMENT

Microsporocytes are closely packed within the anther loculi by four cell layers: the tapetum, middle layer, endothecium and epidermis. At meiosis, mycrosporocytes undergo a typical shape change from polyhedral to roundish, and separate from one another; further, they develop a callose layer interior to the primary wall. A certain number of microsporocytes are subjected to cytological alterations, such as cell deformation and/or cytoplasm degeneration (Chichiriccò 1989a). These are evident after the formation of the callosic wall, and affect a few to many microsporocytes in one or more loculi of the anther. Besides these abnormalities, a number of unimpaired microsporocytes do not complete meiosis. Not only may microsporocyte development fail, but abnormal behaviour of tapetal cells may be concurrent (Chichiriccò 1989a). This tissue is of secretory type (see Pacini 1990), but it shows some tendency to degenerate precociously, as well as to intrude into the loculus. Sometimes, it forms a syncytium around the microspores, similar to amoeboid-type tapetum. Cytokinesis is characterized by the formation of either deformed microspore quartets, or an additional (or incomplete) set of microspores. Dissolution of the callosic wall releases microspores; these show differences in both size and shape (Figure 12.2).

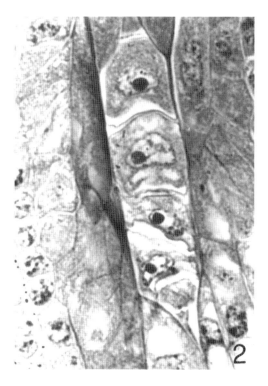

Figure 12.2 Longitudinal section of a saffron crocus ovule showing a pentad of megaspores (× 920).

Microspore development is heterogeneous from both the cytological and structural points of view (Chichiriccò 1989a). The exine wall develops to a standard thickness, not exceeding 0.8 µm; it is a microperforate, colpate and spinulose structure covered with pollenkitt. The underlying intine wall varies from 7.5 to 1 µm in thickness; it consists of two layers, both crossed by tubules, 0.25 to 1.5 nm in diameter, which extend to the exine (Grilli Caiola et al. 1985). The inner layer is notable for its thickenings, protruding into the cytoplasm. Microspores developing through mitosis result in a large vegetative cell and a fusiform generative cell. This bicellular stage is maintained until after pollen dispersion, with sperm cells developing from the generative cell following pollen-tube protrusion. Most pollen grains can hydrolyse starch grains and accumulate lipid globules in the cytoplasm, whereas a lesser number deviate from this developmental programme, accumulating starch grains instead. At anther dehiscence, the size reached by pollen grains varies from 100 to 45 µm; they are roundish, elliptical or cup-shaped. A main distinction may be made on the basis of cytological features: (i) lipoid pollen grains are densely cytoplasmatic (62%), and (ii) starchy pollen grains are poorly represented in cytoplasm (38%). A number of starchy pollen grains include callosic masses which are indicative of cytoplasm disorganization.

Pollen Viability

According to cytochemical tests, most pollen grains from opening anthers exhibit vital activity; however, only a few live pollen grains germinate successfully. Germination may be defective with respect to either the protrusion or growth of the pollen tube. *In vitro*, the most favourable germination has been established in a liquid medium consisting of sucrose and boron (Chichiriccò and Grilli Caiola 1982, 1986), in which 20% of the pollen grains showed germinative activity. The *in vivo* germinability averages 50%, and it persists for several days after pollen dispersion.

MEGASPOROGENESIS AND EMBRYO-SAC DEVELOPMENT

The megasporocyte is, in the ovular primordium, enveloped by parietal tissue and nucellar epidermis. It gives rise, through meiosis, to either tetrads or polyads (Figure 12.1) of megaspores (Chichiriccò 1987). As a rule, the first meiotic division is transverse. The second division may be either transverse or, less frequently, oblique, so the resulting megaspores may be different in both size and shape (Figure 12.1); oblique divisions are recurrent in polyads. During the course of meiosis I, the megasporocyte shows chalazal polarization with regard to starch grains. As a consequence, these are inherited by either the last chalazal megaspore (tetrads), or the two last chalazal megaspores (polyads) (see Chichiriccò 1989b). The embryo sac develops, according to the *Polygonum* type, from the viable chalazal megaspore of the tetrads, while the micropylar ones degenerate. In the polyads, the extra distribution of starch grains, probably associated with other factors, may give vitality to both terminal megaspores. In this case, the embryo sac may arise from the penultimate chalazal megaspore, while the ultimate one may remain living and close to the embryo sac, or both chalazal megaspores may develop, forming two adjacent embryo sacs (Chichiriccò 1987). In 60% of the ovules, the functional megaspore completes three nuclear divisions followed by cellularization, to give a 7-celled embryo sac; this consists of a typical micropylar egg apparatus, a central cell, and three antipodals partly enclosed in the hypostase. Some recurring cytological features in the egg apparatus, such as the synergids with a well-developed, PAS-positive, filiform apparatus and the starchy egg cell, assert functionality of the cellular embryo sacs. In 40% of the ovules, the embryo sac is either nonfunctional, ceasing development at the nucleate stage, or lacking altogether.

OUTBREEDING

The saffron crocus pistil consists of: (i) a trilocular ovary including on average 29 anatropous ovules, (ii) a hollow, 8 to 9 cm long style, and (iii) three 3 to 4 cm long stylar branches forming the stigmas. The stigma rim is provided with papillae; these are unicellular and of the dry type (Grilli Caiola and Chichiriccò 1990), although they lack the proteinaceous pellicle which is typical of dry-type stigmas.

Following pollination, saffron pollen grains stick to the papillae and undergo rapid hydration and germination. Within 50–70 min of pollination, the pollen tubes perforate the papilla cuticle by enzymatic degradation, and grow under it and outside the cell wall, along which a thin layer of exudate is deposited. Below the papillae, the stigmatic pathway for growing pollen tubes lies between the cuticle layer and the inner epidermis of the stigma branches. Along this tract, a copious PAS-positive secretion is released by underlying cells, but a number of pollen tubes fail to grow, the others proceeding to the stylar canal. This is three-channelled, and bordered with elongated secretive cells which discharge polysaccharide-type nutrients into the lumen for the pollen tubes. Along the stylar route, other pollen tubes cease growing, with or without apical anomalies (Chichiriccò and Grilli Caiola 1984, 1986), so few pollen tubes actually reach the ovary. Here, the way to fertilization is along the bottom of the axile grooves, and via placental columns to the ovule micropyle. This tract, based on *Crocus* species, is the main selective site for pollen-tube growth (Chichiriccò 1996, Chichiriccò *et al.* 1995), and is lined with enlarged pyriform cells (Figure 12.3) which, together with the ovule micropyle, secrete flocculent material of a probable glycoprotein nature. Pollen tubes which extend as far as the ovary usually behave like self-pollen tubes of fertile *Crocus* species, ceasing growth in the axile grooves (Figure 12.3). In any event, the fertilized ovules do not succeed in setting seed.

INTERSPECIFIC CROSSES

The pistil of the saffron crocus also supports pollen germination and pollen-tube growth after interspecific pollination (Chichiriccò 1996); crossing partners extend pollen tubes through the stigmas-style to the ovary, just as they do after intraspecific pollination. However, if the partners are not of the *Crocus sativus* aggregate (Mathew

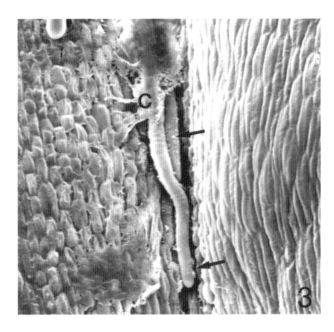

Figure 12.3 Ovarian groove of a saffron crocus under scanning electron microscope showing, under the cuticle layer (C): the pyriform cells of transmitting tissue and, at the bottom, two pollen tubes which have ceased growth; note the flocculent material (arrow) on the pollen-tube wall (bar = 100 μm).

1977), their pollen tubes are rejected along the ovarian grooves, failing to extend to the ovules. Partners of the *C. sativus* aggregate (*C. thomasii* Ten., *C. hadriaticus* Herb., *C. oreocreticus* B.L. Burtt) extend pollen tubes to the ovules and even accomplish fertilization (Figure 12.4) (Chichiriccò 1996). The highest percentage of saffron ovules containing zygote and endosperm nuclei was recovered after crossing with *C. thomasii* (16.8%), followed by crosses with *C. oreocreticus* (11.8%) and *C. hadriaticus* (8.5%). A small number of the fertilized saffron ovules are able to develop to mature seed (Chichiriccò 1989c), according to the embryological pattern of *C. sativus* agg. Abortion frequently occurs before or during the globular stage of embryo development; embryo abortion is frequently preceded by integument degeneration. From 125 saffron flowers crossed with *C. thomasii*, 47 seeds developed to completion. About half of these germinated to plants. The resultant hybrid fruits, seeds and embryos were larger than those of *C. sativus* aggregate.

Interspecific crosses with saffron crocus as the pollen donors were as unsuccessful as intraspecific crosses of saffron crocus (Chichiriccò and Grilli Caiola 1984, 1986).

CONCLUSIONS AND FUTURE PERSPECTIVES

In *Crocus sativus*, the transition from sporophytic to gametophytic generations is characterized by cytological irregularities, most of which are associated with the triploid genome. Therefore, spores are generated which are both genetically and

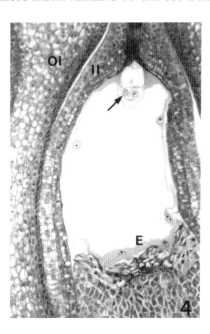

Figure 12.4 Longitudinal section of a saffron crocus ovule 25 days after crossing with *Crocus hadriaticus*. Note nuclear endosperm (E) and the three-celled developing embryo (arrow). OI = outer integument, II = inner integument (× 140).

cytologically unbalanced, and are subject to deviant development. Pollen grains, besides their size and shape, are dissimilar with respect to the development of the intine wall and cytoplasm; on the female side, the embryo sac is deficient in about 40% of the ovules. Together these variations do not account for the total sterility of the saffron crocus. The self-incompatibility mechanisms which prevent inbreeding in fertile *Crocus* species (Chichiriccò 1993, 1996) seem to be also operating in the ovary of the saffron crocus. From a genetic point of view, any pollination event within a population that reproduces only vegetatively is comparable to inbreeding. The reproductive system of the saffron crocus is nevertheless able to support, to some extent, related interspecific crosses. This allows us to obtain seed-producing hybrid saffron plants.

Considering these observations, we conclude that the saffron crocus retains some reproductive-system traits of its presumed ancestors *C. sativus* aggregate, such as the selective role of the ovary, and the potential cross-compatibility, the latter attenuated by the triploid genome producing genetic unbalance. These traits provide promising tools for the genetic improvement of saffron, mostly when associated with methodologies that enhance successful seed set. Integration of the saffron genome with genetic traits from closely related wild species could result in stabilization of traits relevant to breeding; namely, resistance to pathogenic fungi and viruses (Russo *et al.* 1979), induction of hysteranthy (Plessner *et al.* 1989) and improvement of the productivity and quality of saffron drugs (Negbi *et al.* 1989). An appropriate partner for the hybridization programme seems to be *C. thomasii*, a highly fertile and vital species

(Paradies 1957, Chichiriccò 1993) growing in southern Italy and western Yugoslavia. The major drawback of such a hybridization programme is the high number of functional hybrid seeds which need to be tested for the purpose of selection, and the difficulties inherent to natural seed development. Based on previous *in vitro* studies on *Crocus* (Chichiriccò 1990, Chichiriccò and Grilli Caiola 1987), the *in vitro* culture of cross-fertilized saffron ovaries could be a way for successful seed set.

REFERENCES

Brighton, C.A. (1977) Cytology of *Crocus sativus* and its allies (Iridaceae). *Plant Systematics and Evolution*, **128**, 137–157.

Carroll, C.P. (1966) Autopolyploidy and the assortment of chromosomes. *Chromosoma (Berl.)*, **18**, 19–43.

Chichiriccò, G. (1984) Karyotype and meiotic behaviour of the triploid *Crocus sativus* L. *Caryologia*, **37**, 233–239.

Chichiriccò, G. (1987) Megasporogenesis and development of embryo sac in *Crocus sativus* L. *Caryologia*, **40**, 59–69.

Chichiriccò, G. (1989a) Microsporogenesis and pollen development in *Crocus sativus* L. *Caryologia*, **42**, 237–249.

Chichiriccò, G. (1989b) Embryology of *Crocus thomasii* (Iridaceae). *Plant Systematics and Evolution*, **168**, 39–47.

Chichiriccò, G. (1989c) Fertilization of *Crocus sativus* ovules and development of seeds after stigmatic pollination with *Crocus thomasii* (Iridaceae). *Giornale Botanico Italiano*, **123**, 31–37.

Chichiriccò, G. (1990) Fruit and seed development of cultured fertilized ovaries of *Crocus*. *Ann. Bot. (Roma)*, **48**, 87–91.

Chichiriccò, G. (1993) Pregamic and postgamic self-incompatibility systems in *Crocus* (Iridaceae). *Plant Systematics and Evolution*, **185**, 219–227.

Chichiriccò, G. (1996) Intra- and interspecific reproductive barriers in *Crocus* (Iridaceae). *Plant Systematics and Evolution*, **201**, 83–92.

Chichiriccò, G., Aimola, P. and Ragnelli, A.M. (1995) Cytochemical and ultrastructural study of the ovarian transmitting tract of *Crocus* (Iridaceae). *Giornale Botanico Italiano*, **129**(2), 21.

Chichiriccò, G. and Grilli Caiola, M. (1982) Germination and viability of the pollen of *Crocus sativus* L. *Giornale Botanico Italiano*, **116**, 167–173.

Chichiriccò, G. and Grilli Caiola, M. (1984) *Crocus sativus* pollen tube growth in intra- and interspecific pollination. *Caryologia*, **37**, 115–125.

Chichiriccò, G. and Grilli Caiola, M. (1986) *Crocus sativus* pollen germination and pollen tube growth *in vitro* and after intraspesific and interspecific pollination. *Canadian Journal of Botany*, **64**, 2774–2777.

Chichiriccò, G. and Grilli Caiola, M. (1987) *In vitro* development of parthenocarpic fruits of *Crocus sativus* L. *Plant Cell Tissue Organ Culture*, 11, 75–78.

De Nettancourt, D. (1977) *Incompatibility in Angioperms*. Springer, Berlin, Heidelberg, New York.

Ghaffari, S.M. (1986) Cytogenetic studies of cultivated *Crocus sativus* (Iridaceae). *Plant Systematics and Evolution*, **153**, 199–204.

Grilli Caiola, M., Castagnola, M. and Chichiriccò, G. (1985) Ultrastructural study of saffron pollen. *Giornale Botanico Italiano*, **119**, 61–66.

Grilli Caiola, M. and Chichiriccò, G. (1990) Structural organization of the pistil in saffron (*Crocus sativus* L.). *Israel Journal of Botany*, **40**, 199–207.

Karasawa, K. (1933) On the triploidy of *Crocus sativus* L., and its high sterility. *Japanese Journal of Genetics*, **9**, 6–8.

Mathew, B. (1977) *Crocus sativus* and its allies (Iridaceae). *Plant Systematics and Evolution*, **128**, 89–103.

Negbi, M., Dagan, B., Dror, A. and Basker, D. (1989) Growth, flowering, vegetative reproduction and dormancy in the saffron crocus (*Crocus sativus* L.). *Israel Journal of Botany*, **38**, 95–113.

Pacini, E. (1990) Tapetum and microspore function. In Blackmore, S. and Knox, R.B. (Eds.) *Microspores, Evolution and Ontogeny.* Academic Press, London, pp. 213–237.

Paradies, M. (1957) Osservazioni sulla costituzione e ciclo di sviluppo di *Crocus thomasii* Ten. *Nuovo Giornale Botanico Italiano*, **64**(3), 347–367.

Plessner, O., Negbi, M., Ziv, M. and Basker, D. (1989) Effects of temperature on the flowering of the saffron crocus (*Crocus sativus* L.): induction of hysteranthy. *Israel Journal of Botany*, **38**, 1–7.

Russo, M., Martelli, G.P., Cresti, M. and Ciampolini, F. (1979) Bean yellow mosaic virus in saffron. *Phytopathologica Mediterranea*, **18**, 189–101.

13. *IN VITRO* PROPAGATION AND SECONDARY METABOLITE PRODUCTION IN *CROCUS SATIVUS* L.

ORA PLESSNER and MEIRA ZIV[1]

Department of Agricultural Botany, Faculty of Agriculture, Food and Environmental Quality Sciences, The Hebrew University of Jerusalem, P.O. Box 12, Rehovot, 76100, Israel

ABSTRACT   The sterile *Crocus sativus* has a low vegetative propagation rate. Tissue culture techniques were used for rapid propagation of newly introduced varieties and production of pathogen-free corms through organogenesis and somatic embryogenesis. Explants isolated from cauline, foliar and floral tissues were grown on different culture media for the culture initiation stage, the proliferation stage and the hardening, rooting and corming stage. Plant growth regulators, sucrose, active charcoal, coconut milk and ascorbic acid, were employed in the culture media. Callus induced from shoot and corm tissues formed globular embryonic tissues which differentiated into embryoid and matured into plants. Apical and lateral buds regenerated shoots, which developed into microcorms on a 6% sucrose GBR[2] free medium. In the presence of IAA and ZN, embryoids were formed, yet without further development. Corm tissue formed callus and buds which developed into plantlets, or directly into minicorms. Terminal buds exposed to ethylene and to microsurgery resulted in development of axillary buds into microcorms. Floral organs formed, on GBR media, yellow-orange style- and stigma-like structures, in which the levels of the pigments crocin and picrocrocin were 6 and 11 times lower than these in naturally grown stigmas, except in cultures initiated from halved ovaries.

INTRODUCTION

The saffron crocus (*Crocus sativus* L.; Iridaceae), an herbaceous triploid geophyte, is used mainly as a source of secondary metabolites having aromatic and medicinal value. The plant develops annually from buds on the mother corm – a thickened stem, which acts as a resting, perennating storage organ. New corms form via the swelling of the basal internodes of main and axillary shoots. The corms are enclosed by leaves which dry at the end of the growing season, turning into papery scales or tunics (Warburg 1957, Mathew this volume).

---

[1] Corresponding author.
[2] **Abbreviations:** ABA = abscisic ac., IAA = indole-acetic ac., IBA = indole-3-butyric ac., BAP = 6-benzylaminopurine, 2,4-D = 2,4-dichlorophenoxy-acetic ac., GBR = growth bioregulators, GA_3 = gibberellic ac. (gibberellin A_3), KN = kinetin (6-furfurylaminopurine), ZN = zeatin (6-{4-hydroxy-3-methybut-2-enylamino}purine.

Two leaf types develop from the actively growing buds, cataphylls and the true leaves; the former protect the newly emerging leaves, amounting to a total of 12–14 per bud. Adventitious roots, absorbing and contractile, develop at the base of the newly formed corms. The contractile roots assist in anchoring and pulling the daughter corm deeper into the soil.

Crocus sativus, being a sterile triploid plant, is propagated vegetatively by annual replacement corms. Several bacterial, fungal and viral pathogens infest the saffron crocus. These remain active after the corms are harvested and are thus perpetuated. In spite of sanitary measures, the pathogens cause corm and leaf rot, necrosis and often reduce or even inhibit growth and flowering. Plants infested with fungi and bacteria can be treated by bactericidal and fungicidal compounds, while virus-infested plants cannot be treated successfully (Magie and Poe 1982).

As with many bulb and corm plants, meristem-tip culture followed by tissue culture regeneration is almost the only means by which production of clean and pathogen-free propagation material in large numbers can be achieved (Hussey 1975, Debergh and Read 1991). Tissue culture techniques have been applied to the propagation of many geophytes as well as to the saffron crocus. The procedure is based on totipotency – the ability of isolated plant cells, tissue or organs cultured aseptically on a defined medium to regenerate new organs or somatic embryos (Kim and DeHertogh 1996).

Various types of explants have been used for saffron crocus establishment *in vitro*; explants were isolated from the corm tissue, axillary and terminal buds, leaves, nodal tissue and various floral organs (see Table 13.1 for details). The response of the isolated organ or tissue depends on the plant age at the time of isolation, the type of organ isolated and the medium components, in particular the level and combination of plant-growth bioregulators (PBR).

Tissue culture can be a very useful method for effective genetic improvements and production of new *crocus* varieties, because the plant is sterile and conventional breeding methods cannot be used. Protoplast culture, anther culture and the use of various genetic transformations can aid in saffron improvements once *in vitro* methods are established.

MEDIUM COMPOSITION FOR SAFFRON CULTURE *IN VITRO*

Growth and regeneration of *C. sativus* explants *in vitro* have been obtained in both agar and liquid media. The most common ones used are MS, LS, N_6, W and B_5 (Table 13.1). In some of the reported research, any one or all of the mineral constituents are reduced to half-strength level for a better growth response.

The concentration of PBR varies according to the organ used and the culture stage. The medium used for culture initiation – stage I, usually differs from that used for proliferation – stage II, and in many of the cases reported, the medium is changed again for hardening, rooting or corming – stage III.

Growth regulators used for saffron *in vitro* cultures were the auxins NAA, IAA, IBA and 2,4-D, the cytokinins KN, BAP, and ZN, as well as GA_3, ABA, and

Table 13.1 Summary of the *in vitro* studies on *Crocus sativus* L.

Explant Source	Basal Medium[a]	Growth Regulators (μM)	Additives (g per l)	Stage[b]	Morphogenetic Response	References
Lateral buds	MS Macro N_6 Micro and Vit	BAP 10.3 ZN 6.9 GA_3 0.1 NAA 0.05	Sucrose 40	I	Bud growth	Aguero & Tizio (1994)
Lateral buds	MS (1/2 strength in N)	BAP 6.6	Sucrose 30	II	Shoot multiplication	"
In vitro shoots	MS (1/2 strength in N)	—	Sucrose 60	III	Corms, shoots	"
Shoot meristem	MS	BAP 20 NAA 20	—	I	Somatic emryogenesis	Ahuja *et al.* (1994)
Somatic embryos	1\2 MS strength	GA_3 57.8	—	II	Mature embryos	"
Mature embryos	"	BAP 5 NAA 5	Activated charcoal 20	III	Plantlets, corms	"
Floral apices of sprouted corms	MS	2,4-D 9.0 KN 9.3	—	I	Callus	Dhar & Sapru (1993)
Callus	MS	NAA 5.8 KN 9.3	—	II	Corms, shoots flower-like structure	
Small corm sections	MS	KN 2.3	—	I	buds	Ding *et al.* (1979)
"	MS	2, 4-D 4.5	—	II	Cormlets	"
Corms	MS	NAA 5.3 IAA 5.7	—	I, II, III	Callus, plantlets	Ding *et al.* (1981)
Various floral organs	W	NAA 10.7 ZN 11.4	Sucrose 30 Coconut milk 20 Glutamine 0.2	I, II	Stigma-like structure	Fakhrai & Evans (1990)
Petals	W	NAA 21.5–43	"	I	Callus	"
	W	BAP 2.2–35.5 2,4-D 2.2–36.2	"	I	Callus	"
Anthers	W	NAA-43 ZN 22.8		I	Callus	"

[a]MS = Murashige & Skoog (1962); B_5 = Gambourg (1968); LS = Linsmaier & Skoog (1965); N_6 = Nitsch & Nitsch (1969); W = White (1963); N = Nitrogen
[b]Stages of Development: I – Initiation; II – Regeneration and/or proliferation; III – Hardening, rooting, and corm formation

Table 13.1 (continued)

Explant Source	Basal Medium[a]	Growth Regulators (μM)	Additives (g per l)	Stage[b]	Morphogenetic Response	References
Half ovaries	W	NAA 10.7–43.0 ZN 5.6–22.8		I	Stigma-like structure	"
Callus	MS	2,4-D 9.0 KN 2.3	Ascorbic acid 0.1	II	Somatic embryogenesis	"
Globular embryos	1/2 strength MS	IAA 10.7 KN 9.3		II, III	Shoots plantlets	"
Bud-meristem	MS	NAA 21.4 ZN 18.2	"	I	Callus	George et al. (1992)
Globular callus	MS	ABA 3.8	"	II	Shoots	"
Corm and bud section				III	Corm production	Gui et al. (1988)
"	"	IAA 11.4 KN 9.3	Ascorbic acid 0.1	III	Plantlets	"
Various parts of flowers	MS or N$_6$	IAA 11.4 KN 9.3	none	I	Callus	Han & Zhang (1993)
Sections of floral tissue	LS or N$_6$	NAA 26.8 BAP 22.2–26.6			Stigma and style-like structure	Himeno et al. (1988)
Stigma-like structures				I	Synthesis of crocin, picrocrocin, safranal	Himeno & Sano (1987)
Corm fragments	various	NAA 53.7 KN 4.6	Sucrose 20	I	No growth	Homes et al. (1987)
Corm fragments	B$_5$ Minerals, +MS organic	2,4-D 4.5 KN 0.4	–	II	Minicorms	–
Basal parts of leaves	N$_6$	2,4-D 9.0	Same	I	Callus	Huang (1987)
Callus	MS	2,4-D 9.0 BAP 2.2	Sucrose 30	I, II	Buds	"
Buds	1/2 strength MS	NAA 1.1 BAP 2.2	–	I	Shoots	"
Corms	MS	IAA 5.7	Coconut milk 20	I	Callus	Ilahi et al. (1987)
Subcultured shoots	MS	2,4-D 2.3–4.5 BAP 2.2	"	III	Plantlets on the original explant	"

Table 13.1 (continued)

Explant Source	Basal Medium [a]	Growth Regulators (μM)	Additives (g per l)	Stage [b]	Morphogenetic Response	References
Corm	MS	"	"	I	Callus	Isa & Ogasawara (1988)
Callus	liquid MS	2,4-D and ZN	"	II	Small nodules	"
Nodules	MS	2,4-D 4.5	"	II	Enlargement	"
Spherical nodules	MS	NAA 5.5 BAP 4.4	"	III	Shoots	"
Callus culture, stigma-derived	LS	NAA 0.5 BAP 13.3	"		Stigma-like structure	Koyama et al. (1988)
Stalk, young perianth	MS	BAP 30 NAA 50	"	"	Style-stigma-like structure	Lu et al. (1992)
Style, stigma-like structure	MS	KN 3.2 NAA 21.4	"	I, II	Stigma-like structures	"
Various parts of corms, buds, stigmas, peduncles, and leaves	MS	Various	Various	I	Callus from corm segments, shoot and root formation	Milyaeva et al. (1988)
Apical and lateral buds and calli of the explants	MS	IAA or NAA 2.6–16.1 ZN 4.6–13.6	Glucose 30	II	Corms	Milyaeva et al. (1995)
Various parts of flower bud, petal and ovary	LS B$_5$			I, II	Stigma-like structures	Namera et al. (1987)
Flower: connate petal, or ovary	MS	NAA 0.5–5.4 BAP 8.9–22.2 KN 9.3–23.2	Sucrose 50–120		Stigma-like tissue	Otsuka et al. (1992)
Isolated apical buds or buds on small intact corms	MS	ZN 13.7 2,4-D 4.4 ethylene bud explants pre-treatment		I, II, III, III	Leaf and corm production	Plessner et al. (1990)
Young stigmas + ovaries alone or in combination	LS	BAP 0.3, 3.2 KN 4.6, 23.2		I	Stigma growth, crocin biosynthesis	Sano & Himeno et al. (1987)

are narrow, six to seven per corm, and grass-like with an elongated blade reaching a length of 30 cm, produced from October until May. Concomitant with leaf production, growth of roots and daughter-corms takes place. The flowers (up to 12 per corm) bloom before leaf emergence. They consist of violet petals which expand at the top. The pistil consists of an ovary from which a pale yellow, slender style arises and divides into a three-lobed stigma which is orange-red and 2.5 to 3 cm long.

The plant goes through a rest period with regard to active growth in the fall. It also exhibits a short growth period in spring and declines and requires regular and correctly oriented placement.

The most suitable machines for this operation are onion planters, which nevertheless need to be slightly modified, in particular to adapt them to the size of the corms. The machinery tested is of the canopy, three-point linkage type, carried by any tractor, even a light two-wheel-drive tractor. It has a 1.5-m working width and consists of a bulb hopper and a series of scoop wheels which lift the corms out of the hopper and drop them into furrows opened in front by a winged share. The sowing distance along the row is 8–25 cm and 20 cm about 10 cm deep (the planting ratios, however, can be varied). The machine weighs about 400 kg, which requires a tractor having a power of at least 12 hp and, consequently, having possibly a delta suspension to avoid crushing planted corms. Rainfall can be both a delay from 200 to 300 mm.

The working time with this planter per 1000 m² arranged in ridges (four rows per ridge with a total of 55,000 corms) is 5 h, vs. over 100 h for manual planting (Figure 11.1).

Another type of machine that can be adapted to saffron planting is the potato planter: the corms are placed by hand in the scoops, which are moved in horizontal rows by a wheel resting on the ground, by means of chains. The chains are lowered into the ground to deposit the corms in the furrows opened by a furrowing implement located at the front; at the rear, two discs close the furrow.

In the trials, however, only two rows could be planted per ridge, owing to the size of the implements and to the structure, which is intended for only two

Table 11.2 Influence of planting position of corms on the time of emergence and on production

	Normal	Upside Down	In Between
Emergence after 8 days	86.3% control	33.3%	43.2%
Total number of shoots	control	−28.58%	+10.70

¹Other authors have reported a 60% reduction in blossom of tilted corms.

of its production, and (d) the recommended expiry, date for its use. The cooperative makes efforts to find customers in foreign markets and to increase saffron consumption in the internal market.

The product is sold internationally and is delivered by an air-transport agency to the destination port (CIF) from where it is claimed by the recipient following presentation of a Bill of Loading and payment of its value against documents. The transaction is effected by a Greek bank to a foreign bank designated by the buyer.

Because […] every year, and to promote and […] (a) the cooperative re-establish […] in Greek saffron, (b) the state au[…] supporting the crop, (c) the trust of […] funded and (d) cultivation has exp[…]

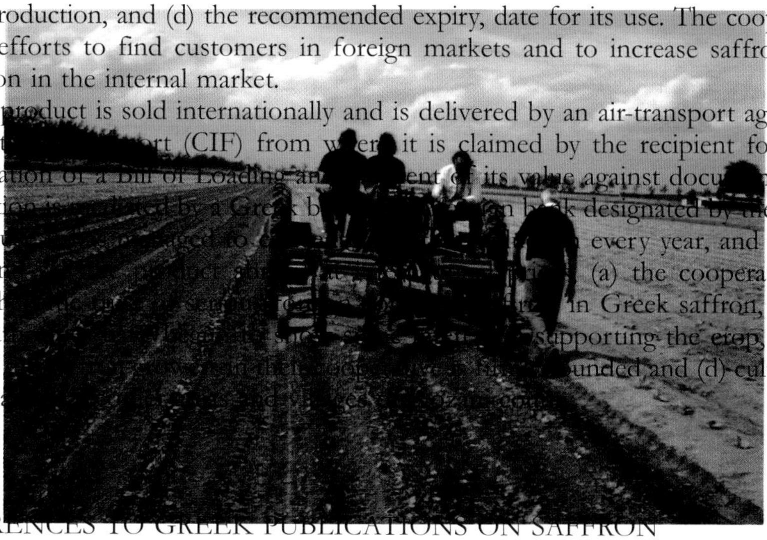

Figure 11.1

REFERENCES TO GREEK PUBLICATIONS ON SAFFRON

Dodopoulos, S. (1977) *Cultivation of Saffron*. Athens.
Goliaris, A. (Unpublished date).
Kritikos, P. (1960) *Crocus*. Athens.
Papanikolaou, A. (1971) *Saffron*. Thessaloniki.
Tahmatzidis, P. (1980) *Crocus of Kozani*. Kozani.

operators. Moreover it was not possible to adjust the distance along the row to less than 150 mm. Overall, this machine was found to give a lower yield than the onion planter but to provide better control over corm orientation (Galigani 1987, Galigani and Adamo 1987, Tammaro 1990).

The potato planter can also be combined with a ridger to prepare and plant in a single operation. This combination produces satisfactory results, reducing the working times (ridging and planting) per 1000 m² to 24 h (8 h/1000 m² for a machine with three operators) (Amato *et al.* 1989).

With regard to trials in Italy, during the 1980s tests were carried out involving burying zinc-mesh cages with a U cross-section (1000 × 80 × 60 mm) containing the corms: this was meant to facilitate the subsequent extraction of the corms from the earth at the end of the cycle (Figure 11.2). This solution was slightly better than the traditional method in terms of the time involved, but the cages were easily damaged, carrying the consequent risk of a considerable increase in costs (Galigani 1987); each cage can be expected to last for 3 years. The working times involved for an area of 1000 m² are reported in Table 11.3. Although the cage system seems to provide the operator with more comfort, it has not met expectations due to the tendency of the cages to warp and of the corms to slide about inside the cages during removal, consequently causing variations in planting density.

WEEDING AND CULTIVATION

The problem of weeds in the first year of cultivation is practically non-existent as blossoms sprout a reasonably short time after having cleared the terrain for the pre-planting stage.

with the addition of coconut milk to 2,4-D and BAP, and followed by bud and plantlet development (Ilahi et al. 1987). A similar response from cultured corms was obtained in the presence of 2,4-D and ZN. The callus redifferentiated into shoots after a period of three months in culture, when ZN was substituted with BAP (Isa and Osagawara 1988).

Small corms were used by Plessner et al. (1991) for shoot development from terminal and lateral buds present on the corms. When the cultured corms were given a pretreatment of exposure to ethylene together with terminal-bud microsurgery, minicorms developed in a medium supplemented with 2,4-D, KN and ZN (see Table 13.2 for details, Figure 13.1). We know of only one report describing the use of leaf explants for callus formation, in the presence of NAA and BAP. The calli produced a large number of buds after a period of eight months and when the medium MS minerals were reduced to half-strength and IAA was added, shoot formation was induced (Huang 1987).

FLORAL ORGANS AS A SOURCE FOR EXPLANTS

The morphogenic response of various floral organs *in vitro*, such as the corolla, ovaries, styles, anthers and stigmas, depends on the age of the flower and stage of development at the time of isolation. In many cases, the callus which forms in the presence of various combinations of growth bioregulators has been observed to redifferentiate into stigma- or style-like structures which contain yellow-orange pigments.

Halved ovaries in the presence of NAA and ZN in the medium (Fakhari and Evans 1990) or whole ovaries or stigmas in a medium containing NAA and BAP (Sarma et al. 1990, 1991) redifferentiated into stigmatic and tubular structures. In one report, as many as 75 stigma-like structures containing yellow-orange pigments

Table 13.2 The effect of ethylene, ethaphon and microsurgery of the apical buds from small corms *in vitro*[a] on axillary bud development after 12 weeks in culture (n-20)

Treatment	Bud Sprouting and Development	Leaves No.
Control (water infiltrated)	1 ± 0	5
Microsurgery[b]	4 corms	–
Ethaphon 1000 mg l^{-1} [c] (infiltrated, 1h)	7±3 basal leaf swelling[d]	–
Ethylene 1000 mg l^{-1} [c] (1h exposure)	6±2 basal leaf swelling[d]	–
Ethylene 1000 mg l^{-1} + microsurgery	15±3 basal leaf swelling[d]	–

[a] The culture medium was supplemented with 2, 4-D 1 mg^{-1}
[b] Wounding of apical buds was carried out with scalpel (n-15)
[c] Ethylene or Ethaphon was administrated as a pre-treatment.
[d] Corms or microcorms developed at the base of apical bud.

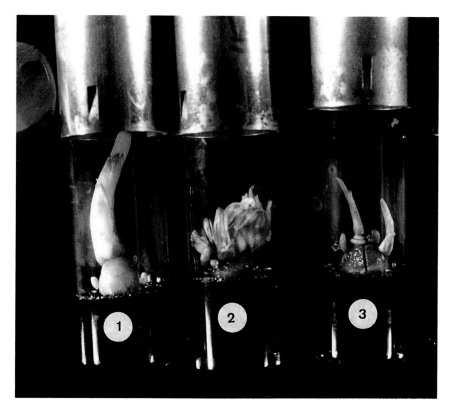

Figure 13.1 Bud development on saffron crocus corms cultured *in vitro* after pretreatment with ethylene or ethaphone. (1) Water control; (2) Ethaphon – 1000 ppm; (3) Ethylene – 1000 ppm (see Table 13.2).

developed in a medium with NAA or IBA (Sano and Himeno 1987). Similarly, whole flowers developed stigma-like structures (Han and Zhang 1993).

Elevated sucrose levels (5–10%) together with BAP, NAA and alanine increased the formation of pigmented stigma-like structures (Otsuka *et al.* 1992). Stigma-like structures on LS medium with NAA and BAP reached up to 15 mm in length after 3–4 months in culture (Koyama *et al.* 1988).

In a SEM study, Himeno *et al.* (1988) showed that the stigma-like structures developed from the cut edge of the ventral epidermal layer of the carpel or hypathium, when cultured in LS medium with NAA and KN at a 10:1 ratio. They found the stigmatic surface of these organs to be mature and biologically functional as a pollen receptor.

SECONDARY METABOLITES IN *C. SATIVUS* IN VITRO

Saffron compounds produced *in vitro* have been reported to develop mainly in floral explants or organs and callus which developed from the floral explants in

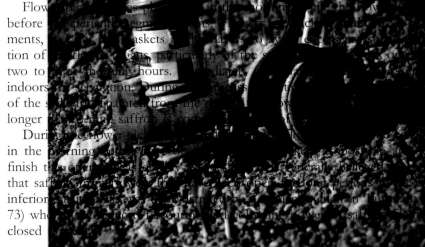

Flowers are harvested in the morning, while they are still closed, before the opening of stigmas. This avoids breakage of the segments, Loading the baskets in thin layers to avoid the deformation of the flower organs, particularly of the stigmas, occurs during the first two to three morning hours. Immediately, the flowers are brought indoors for separation. During this process, stigmas of about 2 mm of the style are separated from the rest of the flower. When the portion is longer, the resulting saffron is considered of inferior quality.

During the flower picking process, all flowers should be picked early in the morning. Usually, more than one person is needed to finish the picking of the flowers in a short time. It is generally believed by farmers that saffron harvested too late in the opening period of the flower yield is of inferior quality. This is confirmed by recent research (Galigani Tuci et al. 1990: 53, 73) which also compared saffron yield and quality between open vs. closed flowers.

Drying of Saffron
Figure 11.3

The fresh red stigmas are dried immediately after harvest. Saffron is handled very gently and carefully to avoid stigma breakage and to ensure optimum conditions for the development of a prime quality product. The stigmas are placed on a cloth in thin layers and dried in the sun for a 2 h period or in the shade after 7 to 10 days. Drying is complete before the stigmas break or crumble. Air-dried saffron retains its purplish red colour, its fragrance and its aroma, and commands a high price in the market place.

Yield

Yield is relatively low in the first year and increases to maximum in the third to fifth years after planting. After this, flower production may decrease. Saffron can vary from 2 to 6 kg per ha, based on planting density, plantation age and climatic conditions during the previous years. The average yield in the country varies from 2 to 2.5 kg per ha, which is very low in comparison to modern saffron plantations in Spain or Italy; lack of rain and irrigation during corm formation and plant growth significantly reduces the yield. One kg of intact flowers yields 72 g of fresh saffron (stigmas), which in turn yields 12 g of dry saffron. The final product retains about 5–20% humidity.

Storage of the Dried Product

In Morocco, saffron is stored as whole dried stigmas and seldom as a powder. Dried saffron is highly hygroscopic; it is kept in well-sealed clay jars or coloured glass containers, or in tightly closed tin cans, and stored in a dark, dry and cool place. It can also be stored in tightly closed dark plastic bags in a dry environment for many years.

SEPARATION OF THE STIGMAS

Average hourly productivity in this activity per person fluctuates between 500 and 1500 blossoms, and the time needed to separate the flowers produced over 1000 m² (about 140,000) therefore varies from 93 to 280 h.

Sorting is always done by hand, even though attempts have been made to separate the stigmas from the stamens and petals by means of a wind tunnel consisting of a variable-section pipe which exposes the cut flowers to an air current. This also happens in flowers gathered by cutting and sucking up, as the two operations are carried out in succession. The use of vibrating boards has not proven suitable either, in separating the stigmas from the stamens and petals (Galigani and Adamo 1987).

DRYING

[text obscured by overlap]

CORM GATHERING

The most common uprooting method is manual, by means of a hoe, on small plots (Figure 8.2). Planting plots are seldom larger than 100 m² and may already be planted with almonds, olives or other fruit trees (Figure 8.3). Otherwise bulb or tuber picking machines may be used. In either case, specific adaptations need to be made. For example, the use of common potato diggers is

---

SAFFRON CULTIVATION IN MOROCCO

annum. Dominant winds blow in a N–NW direction and frost can occur from January to March. Soils in the saffron-growing areas are either sandy loam or calcareous clay with a fairly loose texture. The latter type is dominant in the counties of Taliouine, Zagmouzen and Agadir Melloul.

PROPAGATING MATERIAL

Corm formation and filling occur during the period of vegetative growth (October–March). In March, leaves are cut back and corms undergo a natural dormancy period. During August and September, the corms, which are spherical-shaped and covered with fibrous tunics, are dug up. The daughter corms are separated and subjected to a selection process based on the elimination of rotten, bruised or damaged corms. From the remaining corms, the external 2 to 3 tunics are removed, leaving only the interior one. Only corms with a diameter greater than 2.5 cm are used in propagation. The rest are used as animal feed. Small size corms can result from crowding in the soil, leaf loss or damage before full corm development and maturation, or severe drought during corm formation. Corms can be stored for several weeks in a cool, dry environment, but better sprouting is obtained if they are used shortly after having been dug up (Skrubis 1990).

FREQUENCY OF RENEWAL OF A SAFFRON PRODUCING PLOT

After flowering the parent corm gives rise to two to three new corms which develop to replace the parent. On a given saffron plot, this process continues for several years: every year a corm develops new daughter corms, which can end up crowding each other until insufficient space is left for them to develop to a sufficiently large size to yield a good harvest. Moreover, every year daughter corms usually ascend about 1 cm (5–2 cm) higher in the soil than those of the previous year, until they end up reaching the soil surface. Again, yields decrease significantly and at that stage, the corms have to be uprooted and moved to a new plot. Under the conditions of Taliouine, a given saffron planting can keep producing for 5 to 12 years, on average, depending mainly on the planting density. Higher density plantings give greater yields but need to be replaced within a few years. Between plantings, the field is cultivated for about 3 to 5 years with other crops, mainly cereals, vegetables and legumes.

LAND PREPARATION AND PLANTING

Because of the mountainous terrain, saffron is planted on terraces made on the hill slopes (Figure 8.2). Planting plots are seldom larger than 100 m² and may already be planted with almonds, olives or other fruit trees (Figure 8.3). Before planting, the soil is thoroughly cleared of undesirable material, ploughed to a depth of about 30 cm and then left to rest for several weeks. Ploughing is performed mainly using manual labour and animals.

Koyama, A., Ohmori, Y., Fujioka, N., Miyagawa, H., Yamasaki K. and Kohda H. (1988) Formation of stigma-like structures and pigment in cultured tissues of *Crocus sativus*. *Planta Medica*, **54**, 375–376.

Linsmaier, E. and Skoog, F. (1965) Organic growth factors requirements of tobacco tissue cultures. *Physiologia Plantarum*, **18**, 100–127.

Lu, W.L., Tong, X.R., Zhang, Q. and Gao, W.W. (1992) Study on *in vitro* regeneration of style-stigma like structure in *Crocus sativus* L. *Acta Botanica Sinica*, **34**, 251–252.

Magie, R.O. and Poe, S.L. (1982) Disease and pest associates of bulbs and plants in Koenig, N. and Crowley, W. (eds) *The world of Gladiolus* (N.). North American Gladiolus Council, Edgerton Press, MD, pp. 155–156.

Milyaeva, E.L., Komarova, E.V., Azizbekova, N.Sh., Akhundova, D.D. and Butenko, R.G. (eds). (1988) Features of morphogenesis in *Crocus sativus* in *in vitro* culture. *Biol. kultiviruemykh kletok i biotekhnologiya*, **1**, 146.

Milyaeva, E.L., Azizbekova, N.Sh., Komarova, E.N. and Akhundova, D.D. (1995) *In vitro* formation of regenerant corms of saffron crocus (*Crocus sativus* L.) *Russian Journal of Plant Physiology*, **42**, 112–119.

Murashige, T. and Skoog, F. (1962) A revised medium of rapid growth and bioassay with Tobacco tissue cultures. *Physiologia Plantarum*, **15**, 473–497.

Namera, A., Koyoma, N., Fujioka, K., Yamasaki, H. and Konda, H. (1987) Formation of stigma like structures and pigments in cultured tissues of *Crocus sativus*. *Japanese Journal of Pharmcognosy*, **41**, 260–262.

Nitsch, J.P. and Nitsch, C. (1969) Auxin-dependent growth of excised *Hellianthus* tissues. *American Journal of Botany*, **43**, 839–851.

Otsuka, M., Saimoto, H.S., Murata, Y. and Kawashima, M. (1992). Methods for producing saffron stigma-like tissue. United States Patent. US 5085995,8 pp, A28.08.89 US 399037, P 04.02.92.

Plessner, O., Ziv, M. and Negbi, M. (1990) *In vitro* corm production in the saffron crocus (*Crocus sativus.* L.) *Plant Cell Tissue and Organ Culture*, **20**, 89–94.

Sano, K. and Himeno, H. (1987) *In vitro* proliferation of saffron (*Crocus sativus* L.). *Plant Cell Tissue and Organ Culture*, **11**, 159–166.

Sarma, K.S., Maesato, K., Hara, H.T. and Sonoda, Y. (1990) *In vitro* production of stigma-like structures from stigma explants of *Crocus sativus* L. *Journal of Experimental Botany*, **41**, 745–748.

Sarma, K.S., Sharada, K., Maesato, K., Hara, T. and Sonoda, Y. (1991) Chemical and sensory analysis of saffron produced through tissue cultures of *Crocus sativus*. *Plant Cell Tissue and Organ Culture*, **26**, 11–16.

Sarma, K.S., Ravishankar, G.A., Venkataraman, L.V. and Sreenath, H.L. (1992) Chromosome stability of callus cultures of *Crocus sativus*. *Journal of Spice and Aromatic Crops*, **1**, 157–159.

Visvanath, S., Ravishankar, G.A. and Venkataraman, L.V. (1990) Induction of crocin, crocetin, picrocrocin, and safranal synthesis in callus cultures of saffron: *Crocus sativus* L. *Biotechnology and Applied Biochemistry*, **12**, 336–340.

White, P.R. (1963) *The Cultivation of Animal and Plant Cells*, Ronald Press, New York. **VII**, 228.

Warburg, E.E. (1957) Crocuses. *Endeavor*, **16**, 209–216.

INDEX*

* Ancient authors and books are in bold face

Abscisic acid (ABA)   137–143
Adulteration   45, 47, 95, 97, 98
Aesculapius   73
Alcohol
  absorption   105
  dehydrogenase (ADH$^+$)   35, 38
  detoxification   105
Alexander method   35
Alhagi caamelorum   64
Almonds   89
Alum [$K_2Al_2(SO_4)_2$]   97
Amaranthus blitum   77
Anagalis arvensis   77
Anaphase   128
Anodyne   104
Anthers   58, 139, 144
  culture   138, 139
Anticarcinogenic
  activity   103, 105, 107, 110
Antioxidant activity   107
Antitumor   103, 109, 110
Apiculture   9
Apoptosis   109
Apsheron peninsula   63, 64
Aristophanes   73
Ascorbic acid   140
Atrazine   77
Auxins   137–145
Avena fatua   77

Baking   99
Bees   9, 47
6-Benzylaminopurine (BAP)
    137–145
Bilirbin   104
Biological
  cycle   54
  research   104
Bleaching   98
Blood
  circulation   104, 105
  clots   10

  coagulation   10
  viscosity   104
Bombus silvestris   42
Brain   105
Brassicol   77
Bud   138–141, 145
  culture   143
  meristem   140, 143

Ca^{2+} 42, 43
Cage system   119, 120
Cake saffron   96, 97
Callose   129
Callus   139–142, 144
Cancer chemotherapy   108
Capon juice   99
Cataphylls   138
Capsella bursa pastoris   77
Carcinogenesis   103, 108, 110
Carcinoma   106
Carotene   45–47, 83, 106, 107
Carotenoids   31, 45–47, 99, 107–109
Cell
  central   127
  cycle   68
  membrane   108
  proliferation   108
Cereals   89
Charcoal   142
Chinese medicine   104
Cholesterol   104
Chromatography   8, 47, 49
Chromosome counts   19
Cinnamon   99
Coconut milk   139–141, 144
Coffee grinder   123
Colour   96
Colouring power   92
Corm   5, 6, 31, 55–57, 75–77, 87–89,
    91, 96, 117–120, 123–125, 137, 146
Cormlets   139, 142–144
Corolla   144
Cosmetics   87, 99

assume a certain degree of modernization in saffron cultivation, to move it from traditional agriculture on a small scale carried out by small farmers on marginal land to a more dynamic agriculture where the cost of designing and making specific machines is paid for by the expansion of allotments, and economies of scale may compensate for the inevitable decline in care in the execution of each operation.

This course of action bodes well however, because of the very high production value. Unfortunately, today this value involves costs other than are just as high. Hence labour remuneration turns out to be equal to that of other, less valuable crops, consequently severely limiting the convenience of this cultivation.

REFERENCES

PICKING

Adamo, A., Cozzi, M., Galigani, P.F., Vannucci, D. and Vieri, M. (1987) Fabbisogno di manodopera nelle operazioni colturali dello zafferano. in *Atti Convegno sulla coltivazione delle piante officinali*, Trento 9–10 ottobre 1986, ed. A. Bezzi, pp. 451–452. Istituto Sperimentale per l'Assestamento Forestale e per l'Alpicoltura, Villazzano (Trento).

Amato, A., Amelotti, G., Bianchi, A., Galigani, P.F., Montorfano, P. and Zanzucchi, C. (1989) Zafferano, fonte di reddito alternativo per le zone svantaggiate. *Agricoltura*, 196, 101–128.

Galigani, P.F. (1982) Progetto Piante Officinali: Relazione dell'attività svolta dall'Unità Operativa dell'Istituto di Meccanica Agraria e Meccanizzazione della Facoltà di Agraria dell'Università di Firenze nel quinquennio di ricerca 1978–1982. Pubblicazione speciale.

Galigani, P.F. and Adamo, A. (1987) Le macchine per le officinali. *Terra & Vita*, 10, 62–7.

Langhi, R. (1996) Relazione sull'attività svolta dall'Associazione "Il Croco" di S. Giminiano nell'ambito del programa di ricerche condotte con il contributo della Regione Toscana – 1994–1996.

M.A.F. (1981) Progetto piante officinali: stato della sperimentazione e risultati del primo anno di attività – Ministero dell'Agricoltura e delle Foreste – Istituto Sperimentale per l'Assestamento Forestale e per l'Alpicoltura di Trento.

M.A.F. (1990) Progetto piante officinali: stato della sperimentazione e risultati decennali nel settore delle piante officinali – Ministero dell'Agricoltura e delle Foreste – Istituto Sperimentale per l'Assestamento Forestale e per l'Alpicoltura, Trento.

Skrubis, B. (1990) The cultivation in Greece of *Crocus sativus*. In: *Proceedings of the International Conference on Saffron* (*Crocus sativus* L.), L'Aquila (Italy) October, 27–29 1989, eds. F. Tammaro and L. Marra, pp. 171–182, Università degli Studi dell'Aquila, Accademia Italiana della Cucina, L'Aquila.

Tammaro, F. (1990) *Crocus sativus* L. cv. Piano di Navelli – L'Aquila (L'Aquila saffron): environment, cultivation, morphometric characteristics, active principles, uses. *Proceedings of the International Conference on Saffron* (*Crocus Sativus* L.) L'Aquila (Italy) October, 27–29 1989, eds. F.

12. STERILITY AND PERSPECTIVES FOR GENETIC IMPROVEMENT OF *CROCUS SATIVUS* L.

G. SPARACIO (LOCATED)

Department of Environmental Sciences
University of L'Aquila
Vico Vetoio, 67100 L'Aquila, Italy

ABSTRACT

In *Crocus sativus* L. (saffron crocus), the developmental potential of the spore mother cells is limited by the effect of triploidy, generally linked to a polyploid condition followed especially in response to adverse environmental conditions. However, the saffron crocus, like that of fertile *Crocus* species, supports interspecific crosses with related species. This potential cross-compatibility, together with *in vitro* methods which raise successful seed set, may open the door to breeding programmes for the genetic improvement of *C. sativus*.

INTRODUCTION

Crocus sativus L. is a sterile triploid. Sexual reproduction relies on the beginning of pollen germination at cell level. Fertilization would bring the fertilizing cell through the pistil to the embryo sac. Pollen tube development takes place among young cells through the meiotic process. This comprises a series of coordinated developmental stages of the sporocyte, also correlated with the development of the surrounding sporangium tissues. Any developmental abnormality during meiosis may result in gametophytic sterility. A factor associated with abnormal meiosis is polyploidy. The pollen of polyploid plants, especially triploids, shows a variable degree of pollen sterility (see Carroll 1966). This chapter reviews studies on the reproductive system of the saffron crocus (*Crocus sativus* L.) and its potential with respect to future perspectives for its genetic improvement.

Uses, Marketing and Economics

At the present time, saffron is sold locally to a cooperative or to buyers that serve as intermediaries between the grower and the wholesaler. The cooperative and the buyers sell the product to wholesalers usually located in the main cities of the country such as Casablanca, Rabat, Fès and Marrakech. From there, saffron is either sold in local markets or transferred to other cities. A very small portion is exported, mainly to France and Spain, which in turn re-export some of the quantity to other European countries.

Although the quantity of saffron produced in Taliouine is small, it is a biological product that deserves the highest attention because of its excellent organoleptic qualities. Saffron is highly valued for these qualities and considered a source of income for over 2000 families.

In Morocco, use of saffron is mostly limited to food preparation or to drinks, especially in tea. It adds colour and flavour to all kinds of traditional dishes, to mention only a few such as couscous, tajine, pastilla, harira and pastries.

RESEARCH AND DEVELOPMENT PROGRAMMES

Current yields (2 to 2.5 kg per ha) are very low compared to those obtained in Italy (10–16 kg per ha) or Spain (10–12 kg per ha). Efforts are being devoted to better understanding the factors limiting yield and to improving the existing or introducing new cultural practices (use of fertilizers, selection of plant material, irrigation, etc.). Moreover, to improve the crop's benefit to the grower, organization of the marketing sector is underway, with the creation of a cooperative (with approximately 62% of the total surface area of saffron being only the beginning). This reorganizing, coupled with a good advertising programme locally and overseas, is expected to ensure better marketing and sales. This would bring a better return to the farmers and would contribute to increasing interest in saffron cultivation and agriculture in general among young people.

Experiments to adapt the cultivation of saffron to other areas where its culture might be possible, such as Zagora, located south of Ouarzazate, are being carried out. Other experiments are being carried out in other locations such as Errachidia. Results have not always been conclusive and research continues.

REFERENCES

Agarwal, S. (1987) Saffron – a cash crop of Kashmir. *Agricultural situation in India*, March 1987, 965–968.
Basker, D. and Negbi, M. (1983) The uses of saffron. *Economic Botany*, 37, 228–236.
Basker, D. and Negbi, M. (1985) Crocetin equivalent of saffron extracts. Comparison of three extraction methods. *J. Assoc. Public. Anal*, 23, 65–69.

Medical research   103–113
Medicinal uses   10, 87
Megagametophyte   41
Megaspore   128, 130, 131
Megasporocyte   130
Megasporogenesis   40, 130, 131
Meiosis   127, 128
Meristem culture   138, 142
Metaphase   128
Mice   105, 107
Micropyle   131
Microspore   129
Microsporocyte   129
Microsporogenesis   34, 129
Microsurgery   143
Milling   98
Mineral fertilization   56
Minicorms   140, 143
Mitochondria   35
Mitotic index   66, 68
Moles   77
Mordant   97
Morphology   32
Motor activity   105
Mowing   117, 120
Mulching   120, 121

Necrosis   138
Naphtalene acetic acid (NAA)   138–145
Nematode   91
Nervous system   105
New organs   138
Nodal tissue   138
Nodules   141
Nuclear dimensions   67
Nuclear endosperm   133

Odour   96
Old Testament   73
Olives   89
Onion planter   118, 119
Ontogenesis   65
Organogenesis   65–69, 143
Organoleptic qualities   123

Out breeding   131
Ovary   43, 131, 140–142, 144
Ovule   39, 40, 131–133

Papaver rhoeas   77
Paralysis   104
Pathogens
  bacterial   138
    -free propagation material   138
  fungal   133, 138
  viral   138
Peduncle   5, 141
Penicillium cyclopium   59
Perfume   87
Perianth   140
  tube   5, 6, 70
Petal *see also* corolla   139, 141
Phenology   19, 22–29, 32
Phytoene   47
Phytofluene   45–47
Picking machine (for bulbs and tubers)   123, 124
Picrocrocin   8, 47, 48, 87, 98, 108, 122, 140, 141, 145
Pistil   7, 39–41, 43, 127, 131
Plant growth bioregulators (PBR)   6, 138, 142
Plasmin   10
Plastids   35
Platina   99
Platelet-aggregation   10, 105
Pliny   6, 47, 49
Ploughing   117
Pollen
  development   129, 130
  germination   35, 43, 130, 131
  infertile   41
  grain   7, 35, 37, 38, 130, 131
  micrographs   36
  organization   34
  sterility   127
  tube   7, 35, 127, 130–132
  ultrastructure   38
  viability   34, 35, 130
Pollination   32, 35, 38, 41–43, 131–133

Polyads   130
Polyploidy   127
Potato planter   119
Powdered saffron   45
Propagation   75, 76, 89
Propagation *in vitro*   138, 143, 147
Protocrocine   82
Protoplast culture   138
Pyriform cells   132
Pyrus pyrifolia   9

Quality determination   8

Rats   77
Reproduction   31
Research and development   93
Retinoic acid   108
Rhizocctonia crocorum   77
Ridging   116, 117
RNA   108, 109
Root   6, 7, 56, 138, 141

Safflower   97
Saffron
  analysis   83
  bitter   47
  broth   99
  chemistry   45–52
  cultivation   1–17
  Azerbaijan   1, 63–65, 71
  China   1, 5
  France   1
  Greece   1–4, 8, 75, 77, 78, 80–85
  India   1
  Iran   1, 4
  Israel   1, 2, 31
  Italy   1–5, 8, 9, 31, 53–55, 57, 59–61, 115, 116
  Japan   1, 5
  Mexico   4
  Morocco   3, 4, 87–93
  Libya   4
  Spain   1, 2, 3, 9, 31
  Tibet   5
  Turkey   3–5
  domestication   1

extract   105, 108
gatherer   73
Mechanical cultivation   115–126
packing   98, 99
planting   116, 118, 119
quality   95, 96
tea   96
technology   95, 97, 99, 101
toxicity   10
Safranal   8, 31, 47, 48, 83, 87, 98, 108, 122, 140, 145
Sarcoma   106
Schif's reagent   66
Secondary metabolites   144, 145
Sedative   105
Seed   42, 132, 134
Selection   7, 93
Selenite   107
Shoot   139–143
Simazzine   77
Sinapis arvensis   77
Sleeping   105
Soft drinks   99
Soil analysis   75
Somatic embryogenesis   140
Sonchus oleraceous   77
Sophocles   73
Sperm cells   127
Sporocyte nucleus   127
Sporophytic generation   127, 132
Stalk   141
Stamens   58
Stem apex meristem   66, 67
Sterility   7, 8, 127–135
Stigma   32, 42, 43, 92, 131, 142, 144–146
  culture   2, 5, 10
  drying   8, 9, 58, 79, 81, 96, 97, 123
  -like structure   139–142, 145–146
  separation   79, 81, 91, 92, 96, 117, 123
  supernumerary   8
  storage   8, 9, 93
Style   32, 43, 131, 141, 144
Sucrose   Table 13.1

Microspore development is heterogeneous from both the cytological and structural points of view (Chichiricò 1989a). The exine wall develops to a standard thickness, not exceeding 0.8 μm; it is a microperforate, colpate and spinulose structure covered with pollenkitt. The underlying intine wall varies from 2.5 to 1 μm in thickness; it consists of two layers, both crossed by tubules, 0.25 to 1.5 nm in diameter, which extend to the exine (Grilli Caiola et al. 1985). The inner layer is notable for its thickenings, protruding into the cytoplasm. Microspores developing through mitosis reach in March a tricellular condition. At the anthesis and at the binuclear stage, maintained until after pollinization, biome, with sperm cells developing from the generative cell following pollen tube protrusion. Most pollen grains can hydrolyse starch grains and accumulate lipid globules in the cytoplasm, whereas a lesser number deviate from this developmental programme, accumulating starch grains instead. At anther dehiscence, the size reached by pollen grains varies from 100 to 45 μm; they are roundish, elliptical or cup-shaped. A main distinction may be made on the basis of cytological features: (i) lipid pollen grains are densely cytoplasmatic (62%), and (ii) starchy pollen grains are poorly represented in cytoplasm (38%). A number of starchy pollen grains include callosic masses which are indicative of cytoplasm disorganization.

Pollen Viability

According to a chemical test, most pollen grains during opening anthers exhibit vital activity; however, only a few live pollen grains germinate successfully. Germination may be defective with respect to either the protrusion or growth of pollen tubes. In vitro, the most favourable germination has been established in a liquid medium consisting of sucrose and boron (Chichiricò and Grilli Caiola 1982, 1986), in which 20% of the pollen grains showed germinative activity. The in vivo germinability averages 50%, and it persists for several days after pollen dispersion.

MEGASPOROGENESIS AND EMBRYO-SAC DEVELOPMENT

The megasporocyte is, in the ovular primordium, enveloped by parietal tissue and nucellar epidermis. It gives rise, through meiosis, to either tetrads or polyads (Figure 12.1) of megaspores (Chichiricò 1987). As a rule, the first meiotic division is transverse. The second division may be either transverse or, less frequently, oblique, so the resulting megaspores may be different in both size and shape (Figure 12.1); oblique divisions are recurrent in polyads. During the course of meiosis, the megasporocyte shows chalazal polarization with regard to starch grains. As a consequence, the same is inherited by either the last chalazal megaspore (tetrads) or the two last chalazal megaspores (polyads) (see Chichiricò 1989b). The embryo sac develops, accordingly, from the viable chalazal megaspore of the tetrads, while the micropylar ones degenerate. In the polyads, the extra distribution of starch grains, probably associated with other factors, may give vitality to both terminal megaspores. In this case, the embryo sac may arise from the penultimate chalazal megaspore, while the ultimate one may remain living and close to the

GENETIC IMPROVEMENT OF CROCUS SATIVUS

hay saffron – stigmas in the loose state (*Oxford English Dictionary* 1971) – to completion has been studied (Basker 1993), but use of the resultant product to estimate optimal-drying sensory conditions failed (*ibid*) because, with hindsight, all volatiles had been lost. Proximate analysis of commercial saffron also shows that drying to zero moisture is not an appropriate model (see Table 1 in the chapter on saffron chemistry, this volume). Zarghami (1970) recommends "proper roasting" at unspecified low temperatures. Ameloti and Mannino (1977) write positively of a "fermentation process", [as opposed] to a negative connotation for fermentation during storage (Tammaro 1990)] but do not elucidate. Charcoal fires (Tyler *et al.* 1976; Sokolov 1989; Skrubis 1990; Tammaro 1990) are used for "artificial heating" (Zarghami 1970; Sampathu *et al.* 1984) with few practical details save a remark that "too much heat" destroys the aroma (Ward 1988).

Solar drying, in sun or in shade, has been used as an alternative to "artificial heating" (Watt 1908; Moldenke and Moldenke 1952; Morton and Zallinger 1976; Nauriyal *et al.* 1977; Kapur 1988), even though it is almost guaranteed to result in a photochemical decrease in colour intensity (see the chapter on saffron chemistry, this volume). Drying by solar exposure may be "natural" but as the resultant product shows, it is also crude; the constraint does not apply if sunlight is used only as the heat source, without exposure.

The third drying method, apparently no longer in use, is for the production of cake saffron (Rosengarten 1969; *Oxford English Dictionary* 1971). For this method, a layer of stigmas approximately 6 cm thick was "kiln-dried" under the pressure of a board (Howard 1678; Douglass 1729; *Encyclopaedia Britannica* 1905; Grieve 1959), first for 2 h at one [unstated] temperature, and for a further 24 h at a lower one, also unstated. Our own trials were unsuccessful, possibly because insufficient raw material was available to build up a layer of adequate thickness. The statement that honey and safflower are added to the saffron cake (Wren 1980) must be treated warily: use of honey may be justified technologically as a binder (doubtful), if permitted by local legislation and properly declared on the label; but the addition of safflower could constitute a *prima facie* case of adulteration. The final product has been described as a "compressed matted mass" (Rosengarten 1969).

DYE

Saffron has been used as a yellow dye for wool since ancient times (Basker and Negbi 1983), although Liddell and Scott (1897) feel that this was not yet so in the Homeric era (8th century BCE). A statement that the colour's water-solubility renders it unsuitable for dyeing (Stuart 1979) indicates that the author was unaware that the material to be dyed must be mordanted, *i.e.* soaked in a warm solution of alum [KAl(SO$_4$)$_2$] and cream of tartar [KH tartrate], and then dried (Schetky 1968) before beginning the actual dyeing procedure. The result is a long-lasting yellow. It was once used in Turkish carpets, retaining the colour after decades of constant use (A. Kempinsky 1983, pers. comm.); for economic reasons, saffron is no longer